Making and Using Maps

AGES 5 TO 11

JOHN CORN

Author
John Corn

Designer
Anna Oliwa

Editor
Roanne Charles

Illustrations
David Bowyer

Assistant Editor
Dulcie Booth

Cover image
Alastair Graham

▪ ▪ ▪ ▪ ▪ ▪ ▪ ▪ ▪ ▪ ▪ ▪ ▪ ▪ ▪ ▪ ▪ ▪ ▪ ▪

© 2003 Scholastic Ltd
Text © 2003 John Corn

Designed using Adobe Pagemaker

Published by Scholastic Ltd, Villiers House,
Clarendon Avenue, Leamington Spa,
Warwickshire CV32 5PR

Printed by Bell & Bain Ltd, Glasgow

67890   789012

British Library Cataloguing-in-Publication Data
A catalogue record for this book is available from
the British Library.

ISBN 0-439-98355-X
ISBN 978-0439-98355-6

▪ ▪ ▪ ▪ ▪ ▪ ▪ ▪ ▪ ▪ ▪ ▪ ▪ ▪ ▪ ▪ ▪ ▪ ▪ ▪

The publishers wish to thank:
Ordnance Survey. This publication includes
symbols licensed from Ordnance Survey® with
permission of the Controller of Her Majesty's
Stationery Office © Crown Copyright. All rights
reserved. Licence number 100014536. The symbols
in this publication may be used for educational
purposes only.

CONTENTS

TEACHER'S NOTES

PHOTOCOPIABLES

The photocopiable activities in this book provide a complete scheme of work through which maps and map-making can be taught comprehensively throughout the school. The photocopiable sheets are organised into subject sections and within each section they are ordered progressively, increasing in difficulty to the upper end of primary school. An indication of the age group each sheet is most appropriate for is given in the following teacher's notes. Most of the sheets tie in with the guidance outlined in national documents, either generally or with specific teaching activities. A note of the skills developed in each activity is given under its title.

TEACHER'S NOTES

PLAN FORM

Of all the elements of mapping, children's understanding of plan form – an outline shown from directly above – is one of the most important. It involves representing the three-dimensional world in two-dimensional form on a flat surface such as paper. Many children will not be able to do this until they are perhaps seven or eight, and even then, legs will still appear on desks drawn in plan form. This form is what gives maps and plans their unique character. A plan shows the relative positions of its different parts, for example the rooms of a house, and a map shows the distribution of places. In this book, *plans* are used to describe layouts of buildings such as schools or shopping precincts, while the term *map* refers to larger areas, from a locality to the whole world. It is useful to have a range of plans and maps available for children to use, including plans of the classroom, school and its grounds and maps of different scales showing the local area, UK, Europe and the world. Thematic maps, such as historical maps, land use, rail networks and picture maps (for example, of theme parks) are also very useful.

PICK A PLAN
RECOGNISING PLAN FORM

✳ Before working on the photocopiable sheet, talk to the children about the differences between plan form – seeing an object from above in two-dimensional form and seeing the same object as a picture in three-dimensional form. Using the same object each time, hold it so that the children can see its plan and ask them to describe what they can see. Adjust its position and ask them to describe its picture form. Discuss the differences they find.

✳ Ask the children to draw pictures from 3-D and the corresponding plans of objects around the classroom. Make these into a labelled display.

PAGE
19

5–7

HEROES AND VILLAINS
LOOKING AT SIMPLE PLANS

✳ Encourage the children to look carefully at the outline of roads that the superhero is flying over and decide from the small plan what the buildings are and where on the outline each is located. The children should carefully position the pictures on the outline according to the plan, then stick them on.

PAGE
20

✳ Bring the idea down to classroom level by trying a similar exercise using everyday classroom items scattered over a child's desk. The corners and sides of the desk will provide the locational detail and the children will need to draw and position plans of the objects.

5–7

MOVING IN
INTERPRETING AND USING A PLAN

PAGE 21

✳ Here, the children are asked to use a plan of a small familiar place and make decisions about where to place items of furniture. Ask the children to describe some of the furniture they have in rooms of their own house. Then talk about the layout of the house on the sheet and how 'fixed' features including walls, doors, radiators and windows are shown. Take time to talk to the children about the items of furniture typically found in certain rooms.

✳ Encourage the children to draw a simple labelled plan of a room or series of rooms in their own house.

✳ Help the children to construct a picture plan of the ground and first floor of a 'typical' house. Produce an outline on A3 paper before the children place and stick on items of furniture cut out of a catalogue.

✳ A similar exercise could be undertaken using an outline plan of the classroom and coloured shapes to represent different items of furniture.

5–7

GONE WITH THE WIND
MAKING AND UNDERSTANDING MORE DETAILED PLANS

PAGE 22

✳ This photocopiable sheet shows a mixed-up plan of a school. The children need to cut out the pieces and rearrange them like a jigsaw to make the correct plan of the school. They should then try to name as many parts (rooms) in the school as possible. This activity will help children to appreciate the layout, form and functions of a school building.

✳ The sheet could also be used to introduce work on your own school, where a base plan may be used for the children to record journeys, noisy and quiet places or land use.

✳ Architects' plans can be a useful stimulus and when labelled are good for display.

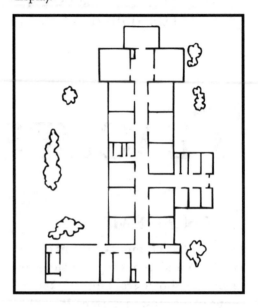

5–9

DESIGN SCHOOL (1) & (2)
DECISION-MAKING THROUGH PLANS

PAGES 23 & 24

✳ Ideally, these sheets should be used to follow on from 'Gone with the wind'. Encourage the children to imagine that they are helping the landscape architect to design the grounds of the new school. Issues such as convenience for children, the use and position of areas of planting and the effects on local people (including those in the retirement home) need to be discussed.

✳ In pairs, ask the children to colour then cut out the features on 'Design school (1)' and try out several different designs on 'Design school (2)' before finally sticking them down in the way they feel is the best. Encourage them to link some of the features with footpaths. Then ask them to write a paragraph describing and explaining their layout.

7–9

TEACHER'S NOTES

MAIN STREET
MAKING MAPS FROM PICTURES

PAGE 25

✳ For this sheet, the children identify plan forms of buildings found along a street. Tell the children to cut out the building plans and stick them on a separate piece of paper in the correct relative locations, so making a low-scale map of Main Street. Each building can then be labelled.

✳ You could draw and photocopy the outline map of Main Street so that the children have an accurate base on which to place the shop plans.

7–9

GREAT SNORING
MAKING MAPS OF SMALL AREAS

PAGE 26

✳ Tell the children to carefully study the illustration showing part of the town of Great Snoring. Ask them to identify the buildings and land uses they can see. They should then make a map of this part of the town on a separate piece of paper and label the physical and human features they have shown. Advise them that the trees, hedges and so on do not need to be shown.

✳ You might want to help some children by drawing the road layout for them to base their maps on.

✳ Children could try a similar exercise using oblique aerial photographs of places nearby. Use a frame to focus attention on a small area of the photograph.

7–9

LITTLE SNORING
INTERPRETING MAPS

PAGE 27

✳ This is an extension of 'Great Snoring', but is more difficult as the children have no information on the height of the buildings to help them complete their sketch. Talk about the relative two-dimensional sizes of the different buildings shown and the shape and location of other significant features. You may want the children to use a frame to choose a smaller part of the map to illustrate.

✳ Use an extract from a local 1:1250 Ordnance Survey map and ask the children to produce a labelled sketch. As in 'Great Snoring', focus attention on a small area of the map by using a frame.

7–9

LAND SURVEY
UNDERSTANDING AND MAPPING LAND USE

PAGE 28

✳ Before completing the photocopiable sheet, brainstorm the different ways that land (or space) is used within school, for example classrooms, halls, offices and corridors. Discuss what a letter code is and why it is used (for speed and space). Explain that a colour code makes the distribution range and amounts of land used for different purposes visually striking and clear. (Suggest to the children that they use different weights or styles of shading if they do not have enough colours.) Only the letter code has been completed in the key, so the children must think carefully about how areas of the school are used. Explain how to rank the different uses.

✳ You could undertake a similar survey, using base plans of your school.

✳ Encourage more able children to draw their own plan of an imaginary school.

The letter codes are: C: classroom, S: storeroom, Cl: cloakroom, Co: corridor, T: toilets, ICT: ICT suite, St: staffroom, L: library, O: office, H: hall, K: kitchen.

7–11

ARCHITECT
DRAWING ACCURATE FLOOR PLANS

PAGE 29

✳ In this activity, the children design the floor plan of one of the houses illustrated on the sheet, using the conventions set out in the on-screen toolbox. It would be useful for the children to see and discuss floor plans of houses (for example, from estate agents) before completing the sheet.

✳ Talk about floor plans in general terms, referring to the conventions shown in the toolbox, and tell the children that each room on their plan should have a window and a door and that there should be two external doors to the house. (Ask them to draw the plan of the ground floor if they choose the two-storey house.)

✳ Display the children's completed plans with the estate agents' sheets.

9–11

LOCATION

Location is another important element of map work and describes where particular features in the landscape can be found. The language of location, for example *near, far, left, right,* is important to establish with young children, predominantly in familiar settings. Older children should begin to develop approaches to locating places through two-, four- then six-figure grid references before they leave primary school. Try to use simple maps of familiar places with younger children, but encourage older children to use more complex OS maps, especially 1:50000.

A GAME OF SOLDIERS
SIMPLE LOCATIONAL LANGUAGE

✳ Using this sheet encourages the children to use locational words such as *left* and *right, up* and *down,* in this case to move a soldier into the correct sentry box. Play some simple games with the children to familiarise them with these words before they tackle the photocopiable sheet.

✳ Other (non-directional) words and phrases can be used to direct the soldiers, such as *past, as far as, up to, turn* and *then.*

5–7

PAGE
30

SAY CHEESE
LOCATING OBJECTS AND/OR PEOPLE

✳ For this activity, the children look at location and its language and a similar activity can easily be undertaken using copies of class photographs. To introduce the sheet, ask the children some simple

PAGE
31

location questions based on your classroom, for example who is sitting to their left or right, nearest to the radiator or your desk. There is a greater emphasis on understanding accurate locations here than in 'A game of soldiers'.

✳ Together, make a plan of your classroom, locating and naming each child on the plan.

5–7

FAMOUS FIVE
DEVELOPING LOCATIONAL LANGUAGE

✳ In this activity, the children are asked to find five easily identifiable characters partially hidden within a busy funfair. Before attempting the activity, ask the children to say where individuals can be found in the classroom. For example, *Gemma is next to the cupboard.* Emphasise the locational words that the children use and brainstorm or suggest some additional ones.

✳ Ask the children to describe the location of other characters in the funfair.

5–7

PAGE
32

SAFARI
INTRODUCING SIMPLE CO-ORDINATES

✳ The map on this sheet introduces the children to simple co-ordinates using two references: a letter (on the x axis), which should be given first, and a number (on the y axis). Ask the children to give the 'best' reference for a certain feature – the reference that includes the most animals or is in the middle of a particular area. Practise giving co-ordinates before attempting to go 'on safari' by using a simple grid drawn on a flip chart.

✳ Ask the children to draw the best route around the park in order to see as many animals as possible, starting and finishing at base camp. Advise them to use grid references in noting the route they take and to mention the different animals that they come across.

5–9

PAGE
33

STAR QUEST
REINFORCING SIMPLE CO-ORDINATES

PAGE 34

✳ This space game will help the children become more familiar with using simple grid references. To introduce it, or for more practice, grids can be placed over picture maps or basic plans, such as of a classroom or school, and the children asked to give letter and number references for individuals or places.

✳ You might want to enlarge the game board onto card to make it more easily readable and more durable. The children will need a counter each and a dice and should play in groups of up to four.

5–9

SHOPPING CENTRE
INTRODUCING FOUR-FIGURE GRID REFERENCES

PAGE 35

✳ This time, the children are encouraged to give more difficult grid references – for places inside a shopping centre in answer to shoppers' questions. Advise the children that they only need to give a reference for one shop even though several may sell the same kind of product, and that they should give a 'best fit' reference, as the grid does not match the outline of the shops.

✳ In pairs, ask the children to make up their own questions and answers.

✳ Try a similar activity, using a plan of a local arcade, supermarket or shopping centre. Place a numbered grid drawn on acetate over the plan before photocopying.

Example answers: 1. 3040 or 2437, 2. 3138, 3. 2537, 4. 2937; 1. 2939, 2. 2738, 3. 2745, 4. 2644 or 2445, 5. 2637 or 2642.

7–9

LOOKING LOCALLY
EXTENDING FOUR-FIGURE GRID REFERENCES

PAGE 36

✳ 'Looking locally' is a development of 'Shopping centre' and asks the children to locate places using four-figure grid references. This activity represents a considerable step in the understanding and use of grid references as the horizontal numbers around the outside of the grid (eastings) and vertical ones (northings) are now marking lines or points on a map rather than areas. The square in the bottom-left corner will be 3143 (and the corner point 310430). Remind them to give 'best fit' references.

✳ It may be useful to introduce this sheet by using a grid drawn on a flip chart with children's names or pictures placed in some of the squares.

✳ Children could give grid references for other features on the map.

✳ Again, grids like this can be put over any local map and the activity repeated.

Answers: 1. 3548, 2. 3545, 3. 4050, 4. 3944, 5. 3848; 1. bowling alley, 2. hotel, 3. video shop, 4. leisure centre, 5. bus depot.

7–9

CRIME SCENE
INTRODUCING SIX-FIGURE GRID REFERENCES

PAGE 37

✳ Explaining how to give six-figure grid references to children can be quite difficult and you might find that considerable introductory work needs to be done on a flip chart or board. The lowest pair of eastings (horizontal numbers) provide the first two numbers in a six-figure reference and the lowest pair of northings give the fourth and fifth. The third and the sixth are each an estimate of the distance from the pairs of numbers to the next, in tenths. So, something at 327472 is seven-tenths of the way from 32 to 33, and two-tenths of the way from 47 to 48.

✳ Remind the children that they are looking for the closest or 'best' reference for a particular clue and to continue on the back of the sheet if they find more.

Possible clues include: finger prints on window frame, bottle and glasses; broken/open window; two glasses; two messages on answerphone; bottle of poison; crumpled paper; bullet case; gun; torn shirt; tickets and passport.

9–11

WHERE'S WHAT?
GIVING ACCURATE SIX-FIGURE GRID REFERENCES FROM MAPS

PAGE 38

✳ Talk to the children about the importance of symbols in everyday life, (see 'Symbols', page 14) and how they are often drawn to resemble the feature they are representing, such as a facility for wheelchair users, a church with a tower.

✳ The photocopiable sheet resembles an OS map at 1:50 000 scale. It will be useful for the children to see a similar map of their local area. Examine the symbols on the map, encouraging the children to suggest what they may represent before checking from the key. To complete the sheet, the children will need to fill in the key before giving accurate grid references.

✳ Undertake similar work using a 1:50 000 OS map.

Answers: 1. 487364, 2. 472318, 3. 512330, 4. 486343, 5. 477378, 6. 474327, 7. 475379, 8. 508327, 9. 468362, 10. 530349.

9–11

WHAT'S WHERE?
FURTHER SIX-FIGURE REFERENCE WORK FROM MAPS

PAGE 39

✳ This is a development of 'Where's what?' with children adding features and symbols to a base map. Talk about the size of the symbols on the map and encourage the children to add theirs at a similar size. Make OS maps available for reference.

✳ For easy marking, put the answers onto tracing paper or acetate and place over the children's sheets when checking accuracy.

9–11

FREE KICK
SIX-FIGURE REFERENCE WORK IN GAMES

PAGE 40

✳ This is a fun approach to giving and plotting six-figure grid references, and can be played in pairs. Give the children a sheet each so both children can shoot and record, but encourage them to work together on one sheet at a time to check the accuracy of the shots.

✳ The children should cut out the football at the side of the goalpost and pierce its centre. One child should look at the goal and defence, locate a place where they believe a goal would be scored if the kick were taken from where the other football is illustrated at the front of the goal area, and note the reference. The other child puts a dot on that reference and centres the cut-out football on it. If the ball would pass cleanly into the net without being stopped by the wall (the defenders) or the keeper (this can be checked by sliding the ball to its new position), it's a goal.

These free kicks will score: 245366 and 250360.

9–11

PENALTY SHOOT-OUT
SIX-FIGURE REFERENCE WORK IN GAMES

PAGE 41

✳ A variation of 'Free kick', this game is another one for pairs of children – one to take the penalty and the other to attempt to save it.

✳ The goalkeepers and ball should be cut out and the centre of the ball pierced and placed on the penalty spot. Explain to the children that the first child in the pair gives a six-figure grid reference to locate a place where they believe a goal would be scored. Then, together, the children put the ball in the goal on that reference. The second player chooses the goalkeeper he or she believes will save the penalty and puts him flat on the page, his 'base' on one of the three positions. Let them know the rule that if any part of the goalkeeper touches the ball, the penalty is saved, if not a goal is scored. Ask the children to take five kicks each and keep a tally as they play.

✳ Children can select their favourite team and colour the goalkeeper's strip accordingly.

9–11

TRAVEL LINES
USING LINES OF LATITUDE AND LONGITUDE

✳ The children can work in pairs on this activity. Before helping them to complete the photocopiable sheet, give them practice in locating places on maps of the world and in atlases, using longitude and latitude references. Explain that lines of latitude are marked in degrees from 0° at the Equator, north to 90° at the North Pole and south to 90° at the South Pole. Lines of longitude are marked up to 180° west and east from 0° at the Greenwich Meridian. Each degree is divided into 60 1-minute intervals. Using one of the examples on the sheet, show the children the convention in co-ordinates of giving the latitude reference first and to 'miss out' degrees and minutes symbols, instead using two spaces between the latitude compass direction and the first number of the longitude reference.

✳ Together, stick resort pictures from holiday brochures onto postcards to make travel agents' cards. Add longitude and latitude references and display them around a wall map of the world.

Answers: 1. Lisbon, 2. Mallorca, 3. Nice, 4. Paris, 5. Geneva, 6. Rome, 7. Barcelona, 8. Tunis.

9–11

GLOBE-TROTTER
ESTIMATING AND CHECKING LINES OF LATITUDE AND LONGITUDE

✳ The children should estimate the references using the lines of longitude and latitude marked on the map and will need to consider lines that would be between those given. Stress that they do not need to include minutes in their answers.

✳ Help the children to compare their references against those on an atlas map. (Indexes may give latitude and longitude references, but this varies between atlases.)

References: Sydney: 34S 151E, Cairo: 30N 31E, Buenos Aries: 35S 58W, Prague: 50N 14E, New Delhi: 29N 77E.

9–11

WORLD CHALLENGE
LATITUDE AND LONGITUDE REINFORCEMENT

✳ This is an extension to 'Globe-trotter' and encourages the children to develop skills in using an atlas to locate places and give accurate references for them.

✳ Allow the children a few minutes to familiarise themselves with atlases, as they are always a great source of interest, and then introduce the photocopiable sheet. Working in pairs, the children should highlight the reference that, after researching from the maps (not the index where references are sometimes printed), they think is correct.

✳ Encourage jackpot-winning children to locate the cities on a blank world map.

Answers: London C, Amsterdam A, Berlin B, Venice D, Jerusalem B, Lagos A, Boston B, Miami D, Tehran C, Karachi C, Lima A, Anchorage D, Jakarta A, Santiago D, Alice Springs B.

9–11

DIRECTION

With young children, the language of direction is similar to that used when talking about location. Direction is concerned with the course or angle that needs to be followed to find a particular feature. After exploring directional language, children should be introduced to four, then (from age 7) eight compass points and have opportunities to develop understanding of what the compass is and how it can be used. Bearings are the most accurate way to give directions and may be tackled by older children.

SCHOOL RUN
GIVING SIMPLE DIRECTIONS

✳ Give simple directions in class and make a list of direction words, such as *up*, *down*, *across*, left and right on a flip chart.

✳ As a class or group, ask the children to describe the journey that they think Katy

Teacher's Notes

makes to school each day. Tick off the words on the flip chart as the children use them. Make a list of the 'landmarks' she passes and put them in the correct sequence.

✳ Encourage the children to do the same for Tom, this time working individually, and write down how he gets to school, paying particular attention to direction words and landmarks.

5–7

STREETWISE
UNDERSTANDING DIRECTIONS USING LANDMARKS

✳ Talk about landmarks and encourage the children to give local examples.

✳ Make a collection of photographs of local landmarks, label and attach them to a large-scale map of the area.

✳ Ask the children to identify each of the buildings shown on the sheet and attach it to the map according to the information in Shani's description.

5–7

WHO AM I?
INTRODUCING THE FOUR MAIN COMPASS POINTS

✳ Look at the four main points of the compass and demonstrate how important they are when giving directions. Practise them using quarter, half and three-quarter turns of the compass face. Together, put up signs in the classroom showing where north, south, east and west are. Use a compass to help.

✳ Establish which children are sitting in each 'direction-area' of the classroom.

✳ Ask the children to fill in the compass points before working on the main part of the sheet.

✳ Encourage the children to play their own 'Who am I?' games on the sheet or on a plan of the classroom labelled with children's names. Some children may be able to use eight compass points.

5–7

ONE WRONG MOVE
DEVELOPING COMPASS POINTS

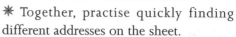

✳ This is a game for two players – one to call out the compass directions in which a character should move to reach the gold, the other to draw the route with a coloured pencil. Remind the children to complete the compass first.

✳ If the first child to take a turn makes a mistake directing a character, the second child tries to get the treasure using the next character (and a different-coloured pencil). When a character reaches the gold, he or she should leave the maze through a different exit. Again, if a mistake is made en route, the first child directs another character, and so on until one manages to leave the maze.

5–7

SANTA SPECIAL
LOOKING AT AREAS USING COMPASS POINTS

✳ This involves some compass work on a plan. Show the children how different houses – detached, semi-detached and terraces – are shown in the plan, and that houses are numbered using even numbers on one side of the street and odd numbers along the other. Point out that most of the houses have not been numbered.

✳ Together, practise quickly finding different addresses on the sheet.

✳ Ask the children to join up diagonally opposite corners of the map and count the number of listed addresses in each of the north, south, east and west triangles created. Note that general directions are considered here rather than specific points.

5–7

RAMBLING ON
INTRODUCING THE EIGHT POINTS OF THE COMPASS

✳ Talk about and practise using the eight points of a compass. To help with this, put signs around the classroom showing each compass point.

✳ Look at the sheet and ask the children to give the compass directions between some of the places. Fill in the compass rose together before letting the children complete the sheet. Advise them to use string pulled straight along the scale to help them work out the lengths of the walks.

✳ Ask the children to create and give directions for other possible rambles. They should trace them on the map and could work out how long each one is.

Directions: Walk 1: N, NW, S, E; Walk 2: E, S, S, SW, NW, NE; Walk 3: W, NW, NW, NE, E, S. The longest walk is Walk 3.

7–9

VAN DRIVER
REINFORCING EIGHT-POINT COMPASS WORK

PAGE **51**

✳ This activity is concerned with giving more complicated directions using an urban street map. Show the children a street map of your local town and ask them to find familiar roads or places on it.

✳ The children should work out the shortest journey as the van travels from store to store, being careful of the one-way system (the dashed lines indicating where the one-way part of the road ends). Each leg of the journey should be traced in a different-coloured pencil and the children's record of the route should detail directions, turns, street names and so on. Stress that the shortest journey in distance is not always the quickest in time.

✳ A similar exercise could be done using copies of local street maps and the locations of real stores and factories.

7–9

COPS AND ROBBERS
EXTENDING EIGHT-POINT COMPASS WORK

PAGE **52**

✳ This is a direction game for two players – one to be the robbers and the other the cops in pursuit. Revise the eight points of the compass and ask the children to complete the compass on the sheet.

✳ Explain that at the start of the game, the robbers are in front of the police car, by ten places in any direction the robbers choose, and the chase starts at the bank. Cars (or counters) move according to the score on a dice. Each player rolls the dice in turn, but the score of the police is increased by one (or two to speed things up) to represent them catching up. When the police are four or fewer places behind, they can radio ahead and employ the 'stinger' – a device for puncturing tyres – to catch the robbers. The stinger is placed five places ahead of the robbers. (This may give the robbers some opportunity to turn off onto another road.)

✳ The police player needs to complete the 'pursuit record' each time he or she rolls the dice, showing road numbers and compass directions. Additional paper (and time!) may be needed to continue playing until the robbers have been caught.

7–9

BEARINGS
INTRODUCING BEARINGS

PAGE **53**

✳ Talk to the children about bearings – what they are, how they are used and how they are more accurate than compass points. (Bearings use all 360 degrees rather than just those at 90, 135 and 180 and so on for E, SE and S.)

✳ Practise estimating then measuring bearings. Construct a sheet of numbered dots spread over centimetre-squared paper, around a central point where a vertical line crosses a horizontal one. Estimate the bearing of each point then place a compass or a 360-degree protractor over the central point to make an accurate measurement. Note the difference between them.

✳ Children should undertake a similar activity on the photocopiable sheet by first estimating then measuring the bearing from Leicester to each of the cities marked.

Bearings: Norwich 78°, London 150°, Cardiff 235°, Birmingham 244°, Belfast 300°, Leeds 339°, Glasgow 325°, Penzance 131°.

9–11

RALLY CROSS
EXTENDING WORK ON BEARINGS

PAGE 54

✶ 'Rally cross' and the following two photocopiable sheets develop the children's understanding of bearings in fun settings.

✶ This is a game for two players. Each player should estimate then state the bearing they wish to travel in (from their current position each time) and enter it on the race card, along with the dice score. The player then moves along this bearing (using a compass to measure it) up to their score in centimetres, even if they crash (which they initially will very often!).

✶ Tell the children to take turns, tracing their journey in different-coloured pencils. Encourage them to have a couple of games, so each player has a chance to start first. If they crash into the ravine or hit another car, a fallen tree and so on, they miss a turn.

✶ Copying the circuit onto centimetre-squared paper could help with accuracy.

9–11

CLAY PIGEON
ESTIMATING BEARINGS ACCURATELY

PAGE 55

✶ Explain clay pigeon shooting to the children to introduce the activity.

✶ Playing in pairs, the children should estimate the bearing from the tip of the gun barrel to the centre of the clay pigeon then draw that bearing on the sheet using a compass or 360-degree protractor. The short guidelines will help with accuracy. The children should take it in turns to shoot from each position and score points according to the accuracy of their estimate.

✶ Add more clay pigeons if required.

9–11

RÉSISTANCE
LOCATING THROUGH INTERSECTING BEARINGS

PAGE 56

✶ 'Résistance' is set in a small French town in wartime. The children use the intersecting points of lines drawn along bearings to pinpoint particular locations. Advise them to use a compass or 360-degree protractor to find the bearings and then draw lines along them. Where a pair of lines cross is an address where a radio transmitter is hidden.

✶ Try some further work using a large-scale – 1:1250 – map, using the example of a TV detector van!

Addresses: 1 Rue de la mairie, 6 Ruelle petite, 23 Rue de Marianne, 2 Rue de Marianne, 5 Rue de nord, 32 Rue de Rouen, bistro – 13 Rue de Rouen, 12 Rue de nord.

9–11

SCALE

Scale is a ratio between the actual distance on the ground and the distance on a map of the same area. It is a method by which we can show part of the Earth's surface on a piece of paper. This is a difficult concept for children (and some adults), who need to be introduced to scale through measuring objects in non-standard then standard measures, before they begin more formal work on scales. When using scales, try to say, for example, that 1 centimetre *represents* or *stands* for 1 kilometre rather than *equals* which is not the case.

LITTLE 'N' LARGE
COMPARING THE SIZES OF OBJECTS

PAGE 57

✶ This activity introduces the concept of scale by encouraging the children to compare the sizes of different animals. In class, compare the heights of different objects and arrange them in order of size. Ask the children to say how big objects are in comparison to others, such as a book to a desk or a child to a teacher.

✶ Show a metre rule to more able children and ask them to use it to estimate the size of different objects and animals, especially those on the photocopiable sheet.

✶ More able children might like to have a go at estimating the sizes of the real animals and positioning the pictures along a scale.

5–7

TEACHER'S NOTES

SIZE WISE
INTRODUCING SCALE THROUGH NON-STANDARD MEASURES

PAGE
58

✳ Let the children measure objects such as books and desks using non-standard measures, for example with pencils or hand-spans, and ask them to record their results. Encourage them to comment about each other's results and suggest why they vary.

✳ Talk about how scale is used to reduce buildings or areas of the Earth's surface so that they can be drawn accurately onto a piece of paper. Explain that the objects on the sheet have been drawn as to-scale versions of their actual size.

✳ When the children have ordered the items on the sheet, again, compare results and discuss any variations.

5–7

CAVENDISH STREET
INTRODUCING SCALE WITH MAPS

PAGE
59

✳ Before asking the children to work on the photocopiable sheet, show them plans and maps of familiar places at different scales. Talk about the areas they cover and how we can calculate the distance between places using the scale given. Remember to use the terms *stands for* or *represents* rather than *equals* when talking about scale, and encourage the children to do the same.

✳ Make a collection of local plans and maps and display them in order of scale.

✳ Use local maps for the same kind of activity as shown on 'Cavendish Street'.

7–9

DISTANCE CHART
LOOKING AT SCALE USING A ROAD ATLAS

PAGE
60

✳ Look at some examples of distance charts and explore how they work. Complicated examples will be found in road atlases; more simple ones in numeracy resource books. Road atlases can be good to use with younger children as they concentrate on making roads and junctions clear and miss out landscape features that would be included on OS maps. Show the children how settlements are shown as dots or shaded areas and point out the different sizes and weight of type.

✳ 30cm lengths of string or strips of paper will be needed so the children can measure the distance between places and then use the scale on the sheet to calculate the actual distance by road. They will need some practice before beginning the sheet.

✳ Make distance charts for use between local places, using a road atlas or OS map. Five to eight places should be sufficient.

Answers: The milometer reads 14588.

DISTANCE CHART (miles)

(shortest distances shown here)

Retford	Lincoln	Newark	Grantham	Sleaford	Market Rasen
28					
25	14				
42	26	18			
38	18	15	14		
26	16	30	42	34	

7–9

VINCE
SCALE USING A ROAD ATLAS

PAGE
61

✳ This is an extension of 'Distance chart'. The children plan Vince's journey and at each stopping point enter the distance travelled in miles and the time taken to get there. Remind the children to include the hour it takes to unload the lorry each time it stops.

✳ In pairs or small groups, children could form their own haulage companies and plan journeys using a road atlas. Give them 'cargo' and 'destination' cards for places in the UK and ask them to record journey times and distances on a chart.

Answers: Glasgow 52155, 9:53; Edinburgh 52195, 11:41; Newcastle 52294, 2:40; Carlisle 52351, 4:48.

9–11

FOR SALE
DRAWING PLANS TO SCALE

PAGE

62

✳ Show the children some estate agents' leaflets and looking at the plans, talk about the rooms and their dimensions and how these are shown on floor plans.

✳ Go through the photocopiable sheet and consider the sizes of rooms (the width dimension is given first) and positions of windows and doors and draw a rough plan on a flip chart. Use the illustration and the starting point marked on the grid (for the front-right corner) to help. The hall is behind the front door and the kitchen beyond that; the living and dining rooms are on the left. Help the children to make their plan accurate and include labels.

✳ Display floor plans and photographs of properties from estate agents' literature.

9–11

TERMINAL 5
USING SCALE ON ATLAS MAPS

PAGE
63

✳ Talk to the children about plane travel, flight times (durations) and time zones. Use an atlas to work out the times of the day in different cities across the world. You may also want to revise the 24-hour clock.

✳ Show the children 'Terminal 5', which shows a simplified time-zones map, and talk through one or two examples, including the general directions. Use the scale to estimate distance and flight time, given an average speed of 600kmph.

✳ Ask more able children to estimate local time on arrival if all flights leave at 12.00.

Answers: Chicago W, 6600km, 11hrs; Mexico City SW, 9000km, 15; Tokyo E, 9500km, 16; Sydney SE, 16800km, 28; New Delhi SE, 6800km, 11hrs 20; Santiago SW, 12000km, 20; Nairobi S, 7000km, 11hrs 40; Moscow E, 2500km, 4hrs 15.

9–11

SYMBOLS

Symbols in mapping are emblems or signs which stand for different physical and human features, and their meanings are usually set down in a key. Help younger children to notice symbols used in everyday life before they look at symbols on a map. Children should begin to understand that symbols are used on maps because they are clear, standard, easily recognisable and take up less space than a written explanation. Colours and shades are also forms of symbol that appear on maps, especially thematic maps like land-use maps, and these can be explored with older children.

WHAT'S MISSING?
INTRODUCING SYMBOLS

PAGE
64

✳ Talk to the children about signs and symbols in everyday life and why they are used. You will be able to point some out around the classroom, such as a fire exit sign. (Show how this also contains a symbol.) Make a collection and display them with captions saying what each symbol means and where it can be found.

✳ Before working on the sheet, look at symbols on different kinds of maps. Ask the children what they think they mean.

5–7

PUFFIN ISLAND
MAKING AND USING BASIC SYMBOLS

PAGE
65

✳ Talk to the children about Puffin Island – its shape, features and settlements. Encourage them to think about the kinds of work that people who live on a small island would do and what would be a clear, easily recognisable symbol for their work and the facilities provided.

✳ Advise the children to complete the key, then add the small symbols to the map to show the facilities available at each settlement.

5–7

EASTWICK FARM
SHADING AS SYMBOLS

PAGE

66

❋ Discuss symbols in more detail with the children and explain how shading can be used to represent and identify buildings or land use on maps. Look at the key on the photocopiable sheet and demonstrate how it can be used to produce a shaded or coloured land-use map of the farm. Explain to the children that you want them to colour the boxes in the key, then use the field numbers to transfer those colours to the map and get a clear indication of how the land is used.

❋ Help the children to produce shaded or colour-coded plans of the school and its grounds, and of other areas close to the school.

7–9

LAXTON
USING LETTER CODES AS SYMBOLS

PAGE

67

❋ Look at the map of Laxton on the photocopiable sheet and ask the children to point to certain features, such as the church, footbridge or school. Go through the key that has been partially completed and ask the children to look for other letter codes on the map. Advise them to think about how land is used in and around the village.

❋ Ask the children to complete the key and map, then link the pictures on the left to the map and draw the features linked to the boxes on the right.

❋ The children could go on to rank the land uses they find, starting with the one that covers the largest area.

❋ For more able children, erase some of the letter codes with correction fluid. This will encourage the children to think carefully about the possible use of some of the land and make decisions about how to record it.

Letter codes: W: woodland, R: river, O: open space, C: church (and churchyard), T: transport, F: farm, H: houses (and gardens), S: school, P: public buildings.

7–9

TOURIST INFORMATION
USING SYMBOLS ON THEMATIC MAPS

PAGE

68

❋ Look at 'Tourist information' together and give the children some quick, simple questions to help familiarise them with the map and the symbols shown. (They could answer these on paper before answering the questions given on the sheet.)

❋ Ask the children to comment on the usefulness of the symbols. If the descriptions are covered, can they say easily what each one stands for?

❋ Look at a local tourist information map. Encourage the children to describe where specific features are and ask them to make a new set of symbols for the map.

9–11

SYMBOLS HUNT
USING SYMBOLS ON ORDNANCE SURVEY MAPS

PAGE

69

❋ Before completing the photocopiable sheet, revise or undertake work on six-figure grid references.

❋ Children should complete the key before answering the questions. Remind them to add symbols of a similar size to those already on the map.

❋ Undertake a similar activity using a 1:50 000 OS map.

9–11

HEIGHT

Height in mapping is a difficult concept for many primary school children as it requires a three-dimensional form to be represented in a two-dimensional way. In this context, height is essentially distance above sea level and on more complex maps is shown through contour lines. These lines allow areas of high and low land to be seen as well as slope and gradient. They are essential when interpreting the landscape or relief of a particular area and a sense of this needs to be conveyed to the older children. Help younger children to

Teacher's notes

understand that land is not flat, that there are hills and valleys. (You may be able to see some from the classroom.) They will know where hills are locally and which are the steepest. The first two activities here will help them to appreciate differences in height and how these can be shown on maps.

High Hopes
Introducing vertical scales

✳ Show the children the illustrations and the scales on the sheet and make sure the children understand that, although the buildings may appear to be of a similar height the scales show that they are not.

✳ Together, estimate the height of one of the Hope family (to the character's eyes) in each of the four places – use the scales and a centimetre ruler to help. Explain to the children how to complete the chart.

✳ Encourage the children to estimate the heights of buildings near your classroom, such as the local church, part of the school, a block of flats. Sketch or photograph each and display them in order of size.

Answers: Simon 20m, Dad 1300m, Flo 140m, Amy 3m, Sarah 9.5m, Rover 900m, Danny 65m, Tim 50m, Mum 1600m, Jack 45m, Tom 6.5m, Kate 100m.

5–9

Highway Code
Looking at gradients

✳ Talk to the children about road signs, especially triangular ones that give warnings to motorists. Explain why it is important for drivers, particularly of HGVs, to be aware of different slopes or gradients and that a high percentage on the sign means a steep slope.

✳ After completing the signs on the sheet, ask the children to devise other ways of informing motorists about the steepness of a hill. Look at a local OS map to see how gradients are shown with black chevrons across the steepest part(s) of a road.

7–9

Hill 'n' dale
Introducing contour lines

✳ Show the children how height is shown on a map and talk about the nature and function of contour lines. (Imaginary lines that join together places of equal height above sea level.) Remind the children about gradient or steepness of slope and demonstrate how this can be seen in the spacing of contours – the nearer together the steeper the slope.

✳ Look at 'Hill 'n' dale'. Ask the children to point to places that are almost level, those that have a gentle slope, moderate slopes and steep slopes, and places that are extremely steep or cliff like.

✳ When describing the walk, encourage the children to include directions, whether it is up- or downhill and features passed on the way.

✳ Identify some other features shown, such as streams, woodlands and crags.

Answers: 1. E, 2. D, 3. F, 4. A, 5. C, 6. B.

9–11

Up and away
Contour layering

✳ Contour layering is a way of recording the height of land so that the distribution of high and low land can be seen easily. Go through the photocopiable sheet with the children and explain the convention of using browns (especially dark brown to represent the highest land), greens, and yellows (to represent the lowest land). You might find it helpful to show the children a relief map of the UK, or part of it, from an atlas that uses this convention. Stress to the children that all of the map on the sheet should be coloured in.

✳ Compare the descriptions of the overhead journey here and those the children wrote for 'Hill 'n' dale'.

✳ Ask the children to comment on the distribution of land levels. They might say, for example, that the highest land is in the north-east and the flattest and lowest land is by the river in a valley bottom.

9–11

PAGE 70
PAGE 71
PAGE 72
PAGE 73

MAP-MAKING

In this section, children are encouraged to put their mapping skills together and begin to draw maps for themselves. Encourage younger children to respond to different kinds of maps and draw maps from experiences and memories that are important to them. Older children can be helped to produce more organised maps that include many, but not necessarily all, of the mapping elements already mentioned in this book. They should also be encouraged to draw different kinds of maps at different scales.

HUFF AND PUFF
MAKING MAPS FROM STORIES

✳ Read the story of the Three Little Pigs from the photocopiable sheet and ask the children to imagine what landscape features there may be in addition to those mentioned in the story. Think about where these features are in relation to each other.

✳ Before the children work on the sheet, demonstrate drawing a picture map of the story, starting with the parents' house and sequencing other features mentioned.

✳ Encourage the children to make labelled imaginary maps based on other stories. Display them under the title of the story.

✳ Ask the children to draw memory maps of their journey to school.

5–7

LOST
DRAWING MAPS FROM SIMPLE INSTRUCTIONS

✳ Ask some of the children to describe their journey to school each morning. Talk about how difficult it can sometimes be following the directions other people give and how easy it is to lose your way. Ask the children whether they have ever been lost – what was it like?

✳ Choose two or three children prior to the lesson and give each of them a self-adhesive label with a letter on it to be stuck somewhere in or close to the classroom. In turn, in front of the class, these children should explain where their letter is to other children whose job, one at a time, is to find each letter. On their return, ask them how easy or difficult they found it.

✳ Go through the photocopiable sheet, and advise the children to read the directions carefully and think about how much they need to put in their map before they begin drawing it.

✳ Ask the children to draw a memory map of their journey to school, then use a local 1:1250 OS map to give directions from school to other local places. Display these directions next to the map and encourage other children to work out where the places are.

5–7

WILD WEST ADVENTURE
MAKING MAPS FROM PICTURES

✳ Look at the photocopiable sheet and discuss the different attractions. Make sure all of the children recognise what they are. On a flip chart or board, invite children to draw some of them in plan form.

✳ The children should cut out each attraction on the sheet and arrange them in the park. Emphasise that they should not stick anything down until they have arranged the whole plan. Tell them to use their plans to draw a map of the park.

✳ Help the children to draw labelled maps of other theme parks or leisure attractions from the picture maps provided in their publicity information.

7–9

ROCKWELL
MAKING MAPS FROM PICTURES

✳ Look at the layout of Rockwell village. Identify the key physical and human features and where they are in relation to each other.

✳ Discuss where the roads should be

positioned to begin drawing the map of Rockwell. Ask the children to map then label key features in the village.

✳ Encourage the children to try to draw other maps from pictures, photographs and small parts of oblique aerial photographs and to label key features.

7–9

MAP-MAKING
ACCURATE MAP-MAKING EXTENSION

✳ This is a version of the more difficult 'plane table radiation' method of map-making which can be used to make a scale map of part of the school grounds.

✳ Before working on the photocopiable sheet, practise the method:

1. Make a cardboard sighter in the shape of an arrow halved lengthways and fix it onto a sheet of paper by attaching a pin to the middle. The pin will let the arrow swivel 360 degrees when the pin is pushed into a clipboard or large flat piece of wood.

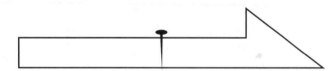

2. Place the equipment outside, on a desk or other flat, stable surface. Using a compass, ensure that the top of the paper and the arrow are pointing towards north.
3. Look along the top of the arrow at nearby features, such as trees or corners of buildings. When a feature is sighted, put a dot on the paper at the tip of the arrow.
4. Write in the name of the feature and measure the distance away using a games tape.

✳ Features below the level of the table, such as flowerbeds or manhole covers, can be sighted using a pole to 'raise' the feature to a visible plane. One child holds a pole at the start and the end of a feature and these positions are sighted, marked on the paper and measured as before.

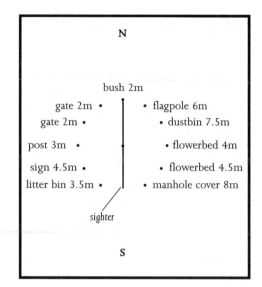

✳ On another piece of paper, help the children to map all the information collected using the scale given. Tell them to use a ruler to join the corners of larger features, such as buildings, and label those features.

9–11

SCHOOL CHASE (1) & (2)
ORIENTEERING GAME

✳ Talk to the children about the sport of orienteering and discuss the 'School chase' photocopiables.

✳ In pairs, children should take it in turns to reach each letter on 'School chase (1)' by estimating a bearing and distance from 'Start' while having a compass or 360-degree protractor nearby. If, after measuring, their estimate is inaccurate they can have two further attempts from the finish position of their first try (not going back to the start) to get closer to and then inside each circle. Only when both players are inside can they collect the letter.

✳ Using the records on 'School chase (2)', ask the players to work out who got round the course in the fewest turns.

The word the letters make is *mapwork*.

9–11

PICK A PLAN

✳ Look carefully at each picture and then choose its plan. Put a large tick ✔ through the correct plan.

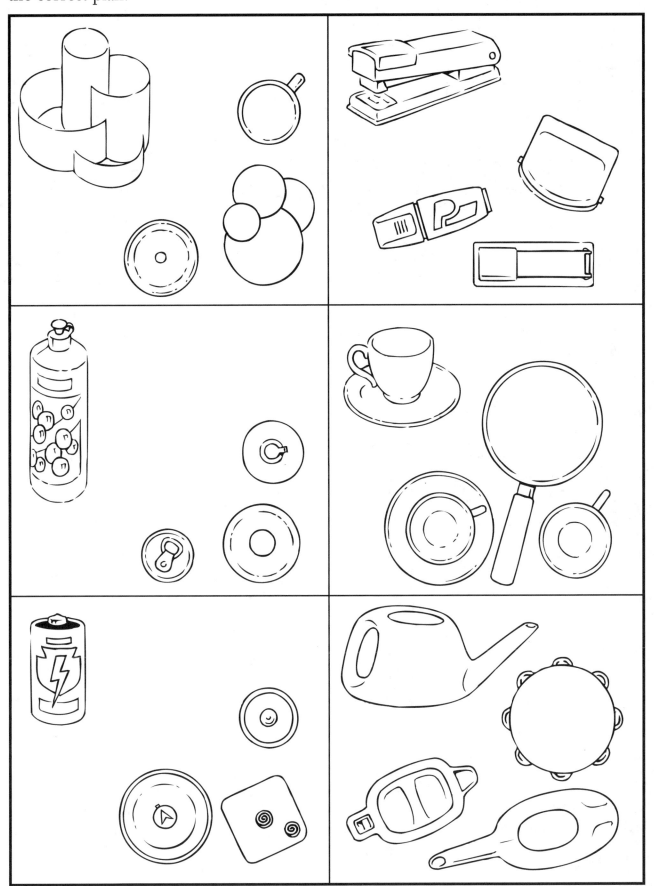

HEROES AND VILLAINS

✺ Use the plan to stick each building in the correct place on the picture.

MOVING IN

✸ Some of your furniture is ready to be moved in. Position the pieces where you think they would fit best.

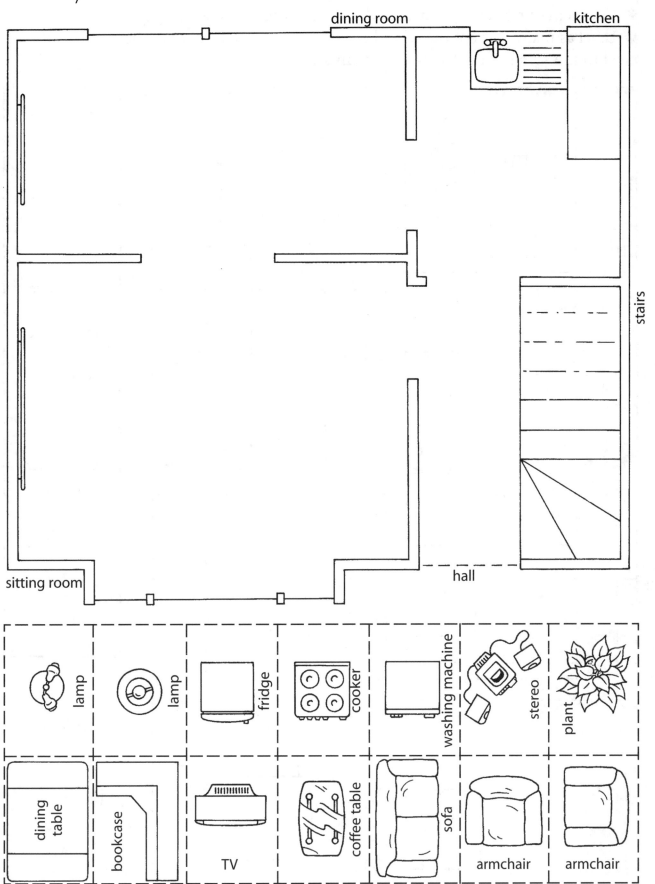

GONE WITH THE WIND

The architect's plans for Windy Lane Primary School blew away. She has managed to get them back, but they are all muddled up.

✳ Cut them out, rearrange them and stick them onto a sheet of paper.

✳ Label as many parts of the school as you can. The list below will help. You will need to use some of the labels more than once.

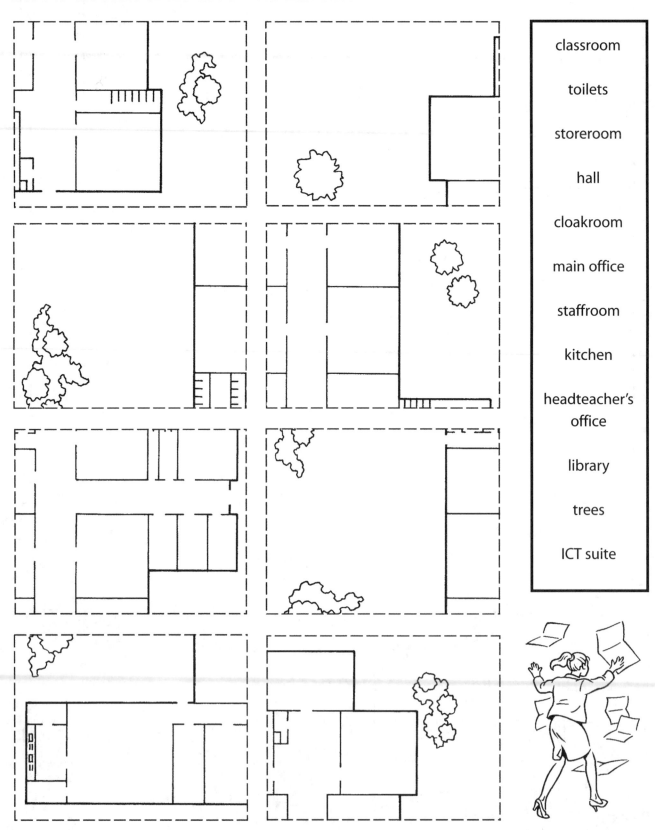

classroom

toilets

storeroom

hall

cloakroom

main office

staffroom

kitchen

headteacher's office

library

trees

ICT suite

Design school (1)

✳ Help the architect finish the school plans. Colour these features for the school grounds and cut them out. Stick them in the best places on the plan and label them.

✳ On a separate sheet, describe where you have put different features and explain why you chose those positions.

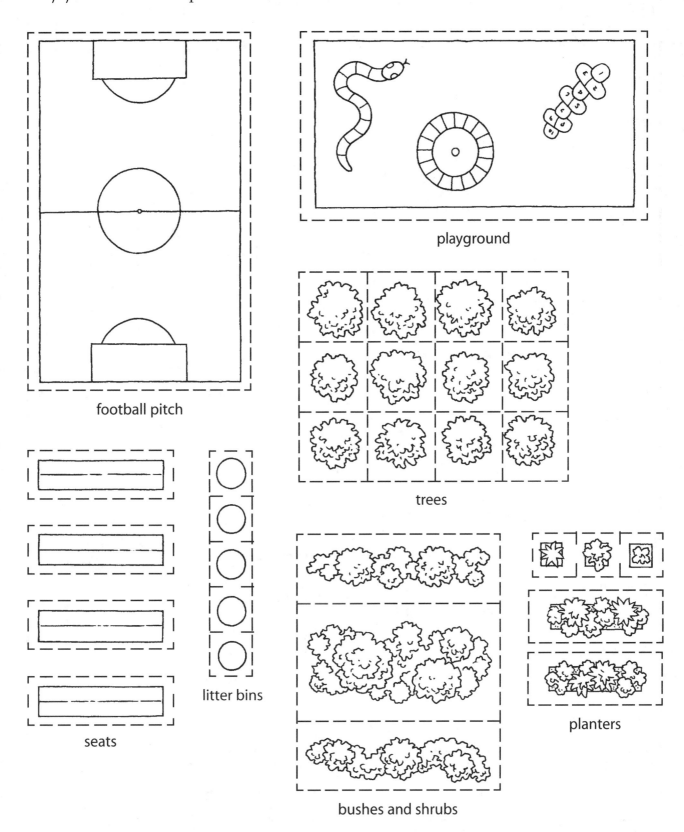

football pitch

playground

trees

seats

litter bins

bushes and shrubs

planters

DESIGN SCHOOL (2)

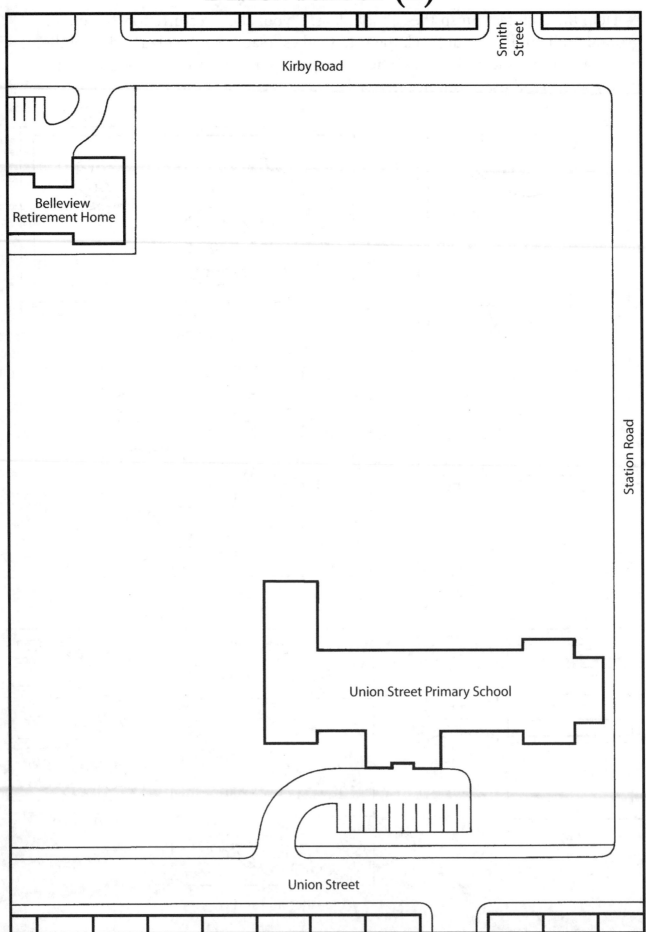

Kirby Road

Smith Street

Belleview
Retirement Home

Station Road

Union Street Primary School

Union Street

SCHOLASTIC TEACHER BOOKSHOP
MAKING AND USING MAPS

📖 SCHOLASTIC PHOTOCOPIABLE

MAIN STREET

* Cut out the plans of the buildings. Use them to make a map of this part of Main Street.
* On your map, label each building and what it is used for.

GREAT SNORING

✹ Look carefully at this picture of Great Snoring and use it to draw an accurate map of this part of town.

✹ Label some of the features you include.

LITTLE SNORING

✳ Look carefully at this map of Little Snoring. Draw a sketch showing what you think the village might look like.

✳ Label some of its features.

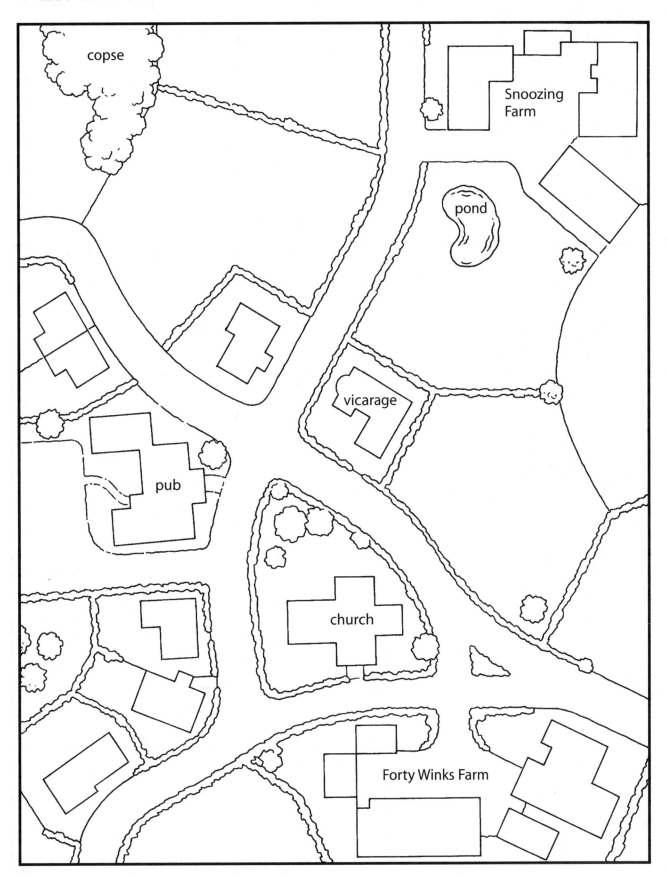

LAND SURVEY

✳ Work out how the land inside Casterley Primary School is used. Use the letter codes to help you. Complete the land uses and colour codes, then colour the plan according to your code.

✳ How is the most land used? How is the least land used? Rank the different land uses in the school. Start with **1** for the way the most land is used.

Land use	Letter code	Colour code	Rank
classroom	C		
	S		
	Cl		
	Co		
	T		
	ICT		
	St		
	L		
	O		
	H		
	K		

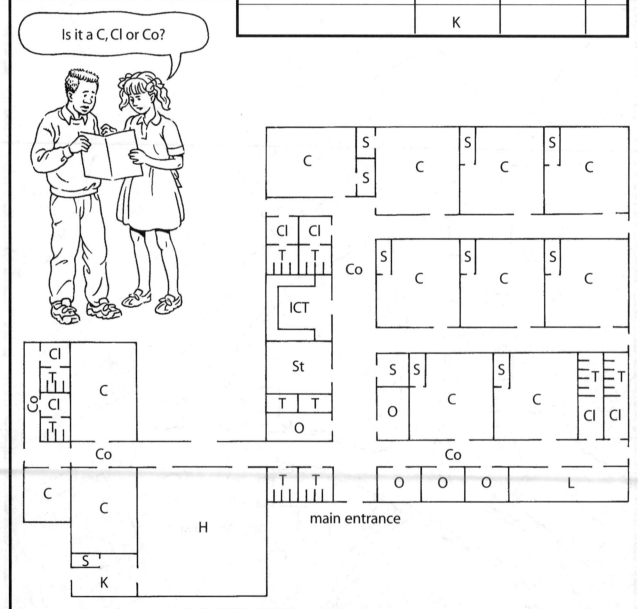

Is it a C, Cl or Co?

main entrance

ARCHITECT

COMPUTER AIDED DESIGN

EXIT

TOOLBOX
START

— wall
- - - window
door

stairs

bushes

tree

24.02

✳ The architect has to design a floor plan for one of these two houses. Choose a house, and use the toolbox to draw and label its floor plan.

A GAME OF SOLDIERS

✴ Take the soldiers back to their sentry boxes.

✴ Write down how you did it. You may find some words in the wordbank helpful.

Wordbank
past
crossroads
turn
when
after
up
down
forward
then
along
straight on

1 _____

_____

2 _____

_____

3 _____

_____

SAY CHEESE

Class 1b Summer

✳ Look carefully at the summer photograph of Class 1b.

1. Who is third from the left on the back row? _____

2. Which child is two places to the right of Jade? _____

3. Which child is directly behind Tom? _____

4. Who is five places to the left of Suki? _____

5. Who is in front of Kate then two places to the right? _____

6. Which child is in front of Dulcie? _____

7. Where is Joe? _____

8. Where is Laura? _____

FAMOUS FIVE

✸ Find these five people and describe where you found them.

1.   2.   3.   4.   5.

SAFARI

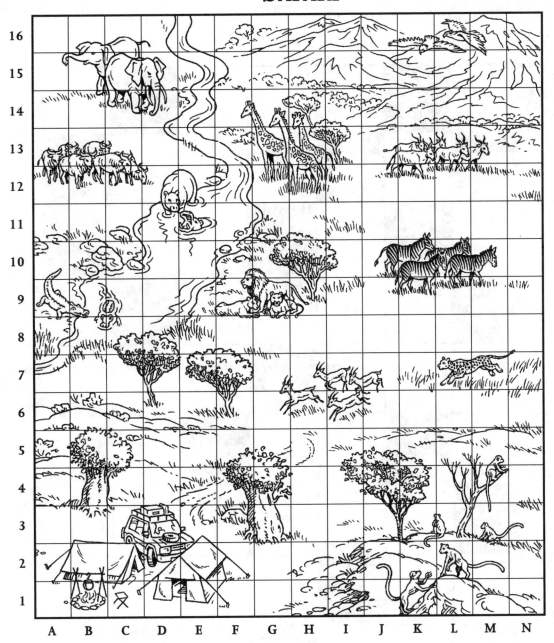

✳ Where can you find…?

1. zebra _____

2. lions _____

3. hippos _____

4. elephants _____

5. monkeys _____

6. giraffe _____

7. eagles _____

✴ Using co-ordinates, describe a journey from the base camp and back, seeing as much wildlife as you can. (Continue on the back of the sheet if necessary.)

STAR QUEST

✳ Your mission is to take the valuable cargo of pluranium to Space Dock 2.
Good luck!

	A	B	C	D	E	F	G	H	I
12	FINISH — Space Dock 2		Dark Matter — Give reference for Keni Alpha to pass				Langon trouble — Go to D11		◄
11	►	Ion speed 6 — Go to I11		Langon				Wormhole — Go to F7	
10			Observe supernova — Go to A10		Slow to ion speed 2 — Miss a turn			Wormhole — Go to F12	◄
9	►	Attack by Xenoshaker — Detour to G8					Black hole — You lose!		
8					Ion drive fails — Return to I8				Stop at space station — Return to Keni Alpha ◄
7	► Keni Alpha		Rescue Alalidds — Return to B4				Wormhole — Return to C4		
6		Black hole — You lose!							Asteroids! — Back to H5 ◄
5	Fight Creepies ► Go to F4			Empty quadrant — Go to H7					
4		Space station			Avoid comet — Go to I4		Time loop — Miss 3 turns		◄
3	Goth attack ► Go back to E2		Wormhole — Go to H6					Space dust — Go to G3	
2					Ion speed 5 — Go to G3		Pod problems — Return to A1		◄
1	Space Dock 1 — START ▶		Ion speed 2 — Go to F1			Solar wind — Go to E2			

SHOPPING CENTRE

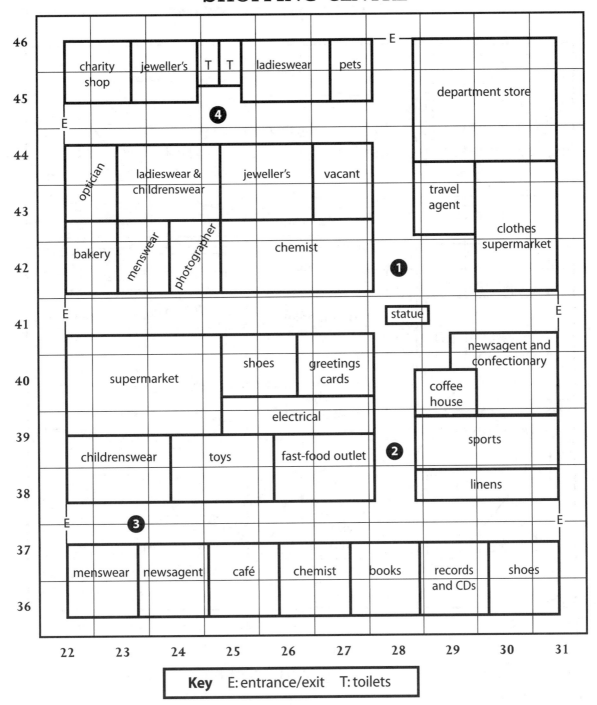

Key E: entrance/exit T: toilets

✳ Give the correct grid reference to help these people. Their current positions are numbered on the plan.

❶ Where can I buy a magazine? _____

❷ Where is the nearest exit? _____

❸ Where can I buy a cup of tea? _____

❹ Where can I get a CD? _____

✳ Give the reference for a shop where you can buy:

1. a tennis racket _____

2. a burger _____

3. a guinea pig _____

4. a watch _____

5. a packet of aspirin _____

LOOKING LOCALLY

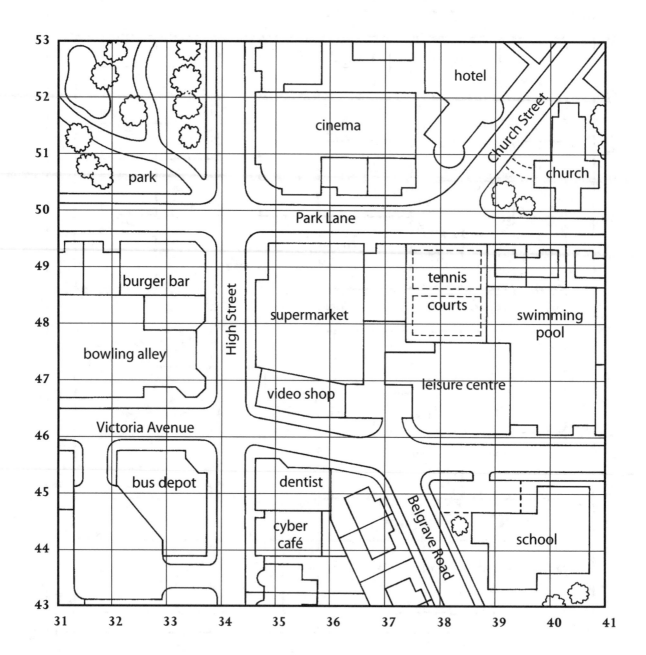

* Give the best four-figure references for these places:

1. supermarket _____

2. dentist _____

3. church _____

4. school _____

5. tennis courts _____

* What will I find at or near:

1. 3147 _____

2. 3852 _____

3. 3546 _____

4. 3846 _____

5. 3344 _____

CRIME SCENE

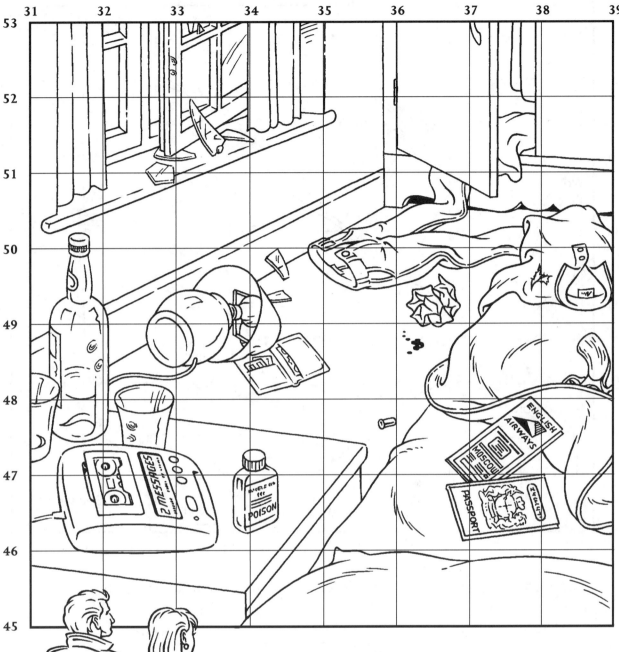

✸ Help the officers investigate the crime scene.
Find more clues and note their positions with
six-figure grid references.

CRIME SCENE REPORT	
Clue	**Reference**
fingerprint on window	330524

WHERE'S WHAT?

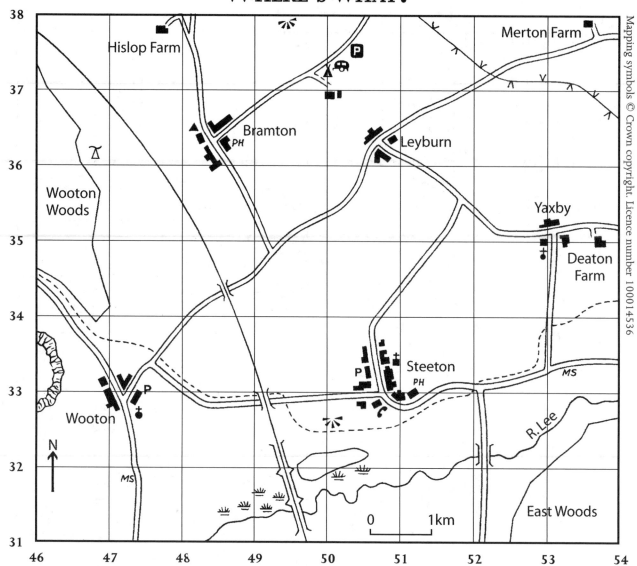

❊ Use an Ordnance Survey map to complete this key.

PH	_____
☐	car park
☐	milestone
⌣	_____
☐	viaduct
⌄	_____
☐	footpath
▲🚐	_____
🎇	_____
☐ ☐	church with spire, with tower

❊ Give the most accurate six-figure grid reference you can for:

1. pub in Bramton ___487364___

2. milestone near Wooton _____

3. pub in Steeton _____

4. bridge to the south of Bramton _____

5. Hislop Farm _____

6. church in Wooton _____

7. viewpoint NW of Leyburn _____

8. telephone kiosk in Steeton _____

9. radio mast near Wooton Woods _____

10. church in Yaxby _____

Mapping symbols © Crown copyright. Licence number 100014536

WHAT'S WHERE?

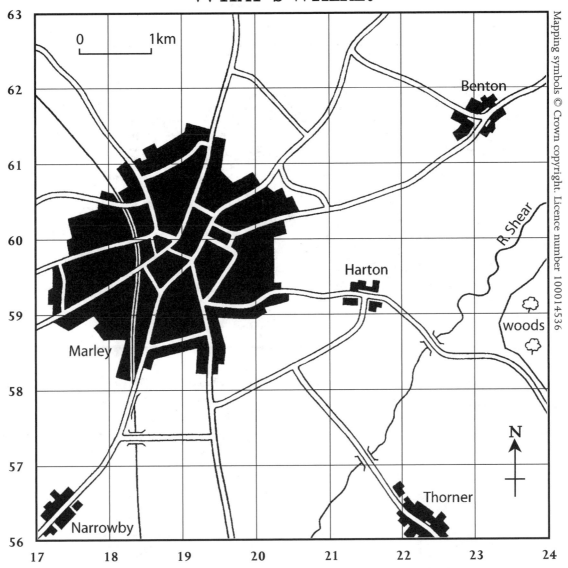

Mapping symbols © Crown copyright. Licence number 100014536

* Help this worried map-maker finish the map. Put in the following features:
* post offices at **219562** and **233615**
* churches with towers at **192582** and **197609**
* churches with spires at **175567** and **188614**
* a new road between **194574** and **217567**
* pubs at **213592** and **171567**
* a viewpoint at **197629**
* milestones at **239621** and **172624**
* a footpath between **219630**, **215615** and **240610**

FREE KICK

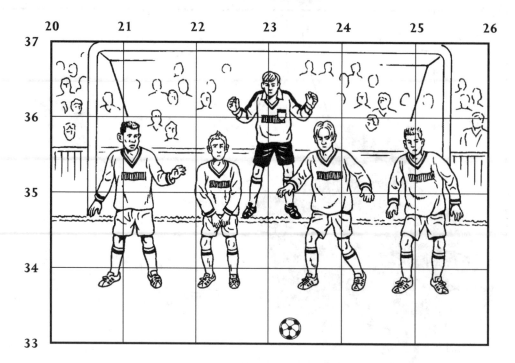

✳ Take five kicks. For each, give the six-figure grid reference which you think would send the ball into the net.

✳ Tick which of these free kicks will score.

223363	245366	214359	230345	250360

PENALTY SHOOT-OUT

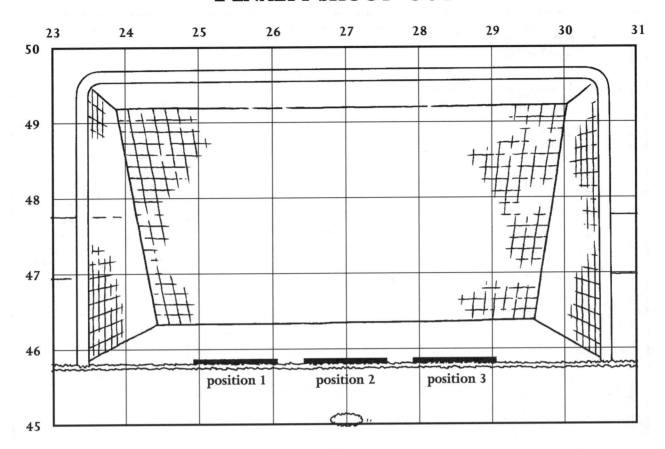

	Reference	Goal	Saved
Kick 1			
Kick 2			
Kick 3			
Kick 4			
Kick 5			

✹ You have five penalties to take. Your teacher will tell you how to play. Mark your attempts as six-figure grid references and remember to tick whether each penalty is scored or saved.

TRAVEL LINES

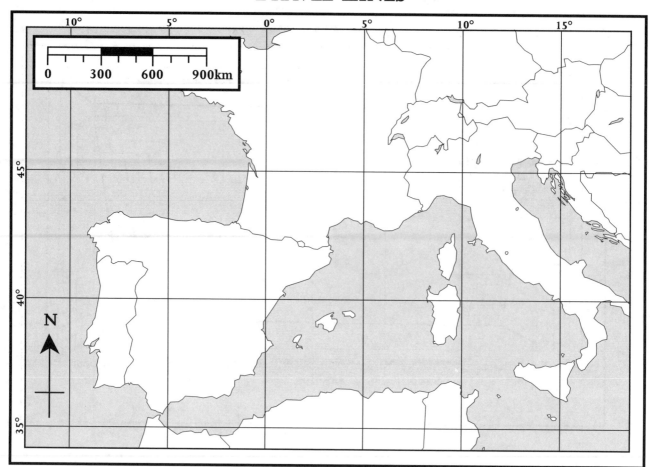

❋ Use an atlas to help you complete the cards in the Travel Lines window below.
❋ Mark the holiday destinations on the map above.

1. 3-star hotel in Europe's most western capital city only £389

_____

(38 42N 9 10W)

2. 2 weeks half-board in the sun £419

_____

(39 30N 3 0E)

3. Experience the glitz and glamour of

_____

7 nights £525
(43 42N 7 14E)

4. Romantic capital city escape £199

_____

(48 50N 2 20E)

5. Lakes and mountains £485 May/June

_____

(46 12N 6 9E)

6. 7 nights 4-star city hotel only £425

_____

(41 54N 12 30E)

7. Spain old and new. Weekend break £259

_____

(41 21N 2 10E)

8. On the edge of the Sahara. 2 weeks full board £649

_____

(36 50N 10 11E)

GLOBE-TROTTER

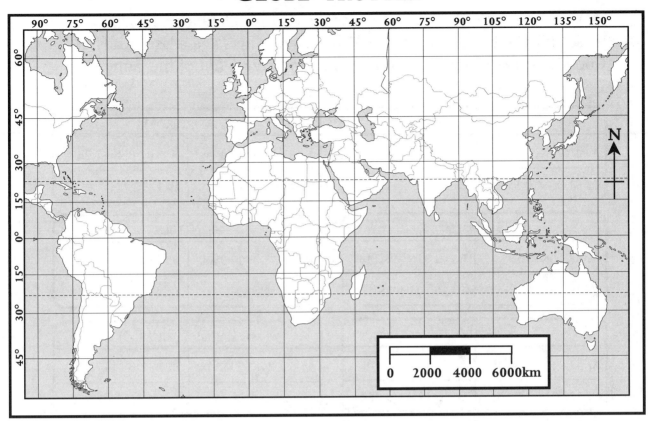

❋ Mark on the map each city shown in the passport. Use an atlas to help.

❋ For each city, add your estimate of its latitude and longitude reference.

❋ When you have finished, use an atlas to check your references.

WORLD CHALLENGE

❋ Welcome to *World Challenge*. Choose the correct reference for each of these cities and you'll win the jackpot! Highlighters and atlases ready… Good luck. Remember – get one wrong and you're out of the game.

1. LONDON	
A. 007	**B.** 999
C. 51 30N 0 5W	**D.** 3½

2. AMSTERDAM	
A. 52 23N 4 54E	**B.** 100 0N 55 0E
C. 2, 4, 6, 8	**D.** 24–7

3. BERLIN	
A. 68 10N 103 50E	**B.** 52 32N 13 24E
C. 200200	**D.** 50 10S 99 9E

4. VENICE	
A. 3 15N 100 0W	**B.** 9876543
C. 15 10S 83 17W	**D.** 45 27N 12 20E

5. JERUSALEM	
A. 52 36N 1 59W	**B.** 31 47N 35 10E
C. 38 52N 77 0W	**D.** 19 57N 79 11E

6. LAGOS (NIGERIA)	
A. 6 25N 3 27E	**B.** 43 0N 44 35E
C. 57 43N 16 43E	**D.** 36 40N 48 35E

7. BOSTON (USA)	
A. 33 2S 137 30E	**B.** 42 20N 71 0W
C. 37 25N 139 28E	**D.** 20 2N 74 30E

8. MIAMI	
A. 45 48N 3 4E	**B.** 30 45S 25 5E
C. 43 48N 78 10W	**D.** 25 52N 80 15W

9. TEHRAN	
A. 28 8N 84 40E	**B.** 16 30N 107 35E
C. 35 44N 51 30E	**D.** 19 20N 81 20W

10. KARACHI	
A. 28 38N 72 17E	**B.** 43 21N 42 30E
C. 24 53N 67 0E	**D.** 35 44N 51 30E

11. LIMA (PERU)	
A. 12 0S 77 0W	**B.** 0 15S 78 35W
C. 4 34N 74 0W	**D.** 26 21S 31 52E

12. ANCHORAGE	
A. 61 0N 160 0W	**B.** 61 0N 160 0W
C. 49 51N 120 10W	**D.** 61 10N 149 50W

13. JAKARTA	
A. 6 9S 106 49E	**B.** 14 40N 120 35E
C. 1 17N 103 51E	**D.** 21 11N 114 14E

14. SANTIAGO (CHILE)	
A. 37 14S 7315W	**B.** 30S 25W
C. 36 17S 70 40W	**D.** 33 24S 70 50W

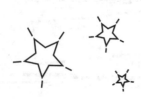

15. ALICE SPRINGS	
A. 1810S 175 10E	**B.** 23 36S 133 53E
C. 17 0S 176 50E	**D.** 16 10S 178 30W

❋ How far did you get?

School run

✳ Katy and Tom both come to school by car. Use different colours to mark how they get there.

✳ On a separate sheet, describe Tom's journey.

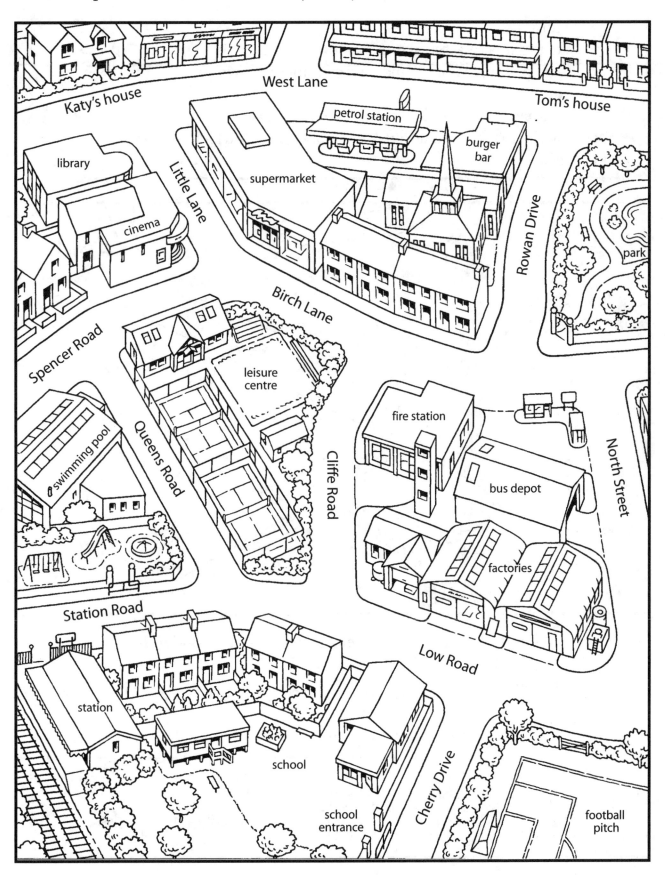

STREETWISE

When I walk to school,

* I pass the sweet shop opposite, on the corner of John Street.
* Between John Street and Elm Street, there is a supermarket, then a chemist, then a cinema.
* Opposite the supermarket, across London Road, there is a church.
* After that I pass a post office and a row of houses down to Kings Road.
* Between Kings Road and the school there is a garage.
* Opposite the garage are a hospital and a big car park which reaches Wagon Lane.

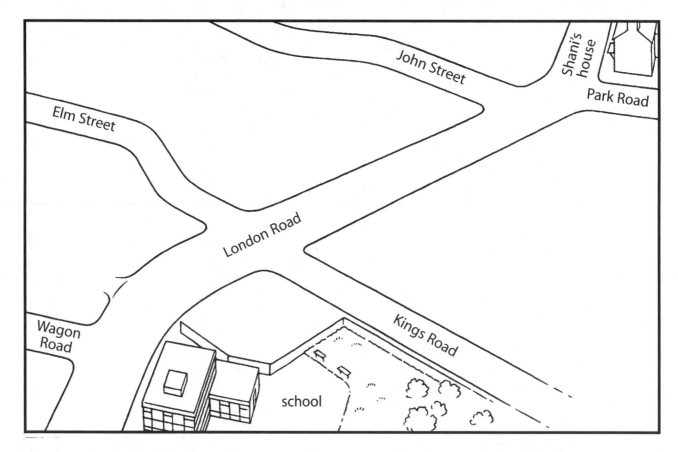

* Cut out the buildings and position them correctly along London Road.

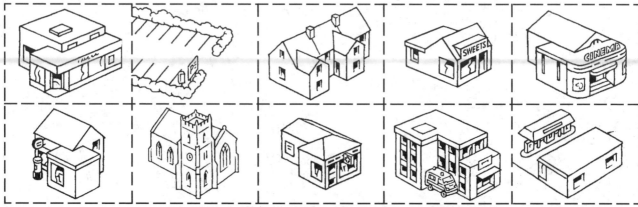

SCHOLASTIC PHOTOCOPIABLE

WHO AM I?

* Start at Daniel. Go south 2 children and west 2 children. Who am I?

_____

* Start at Rhian. Go north 3 children, east 2 children and south 2 children. Who am I?

_____

* Start at Matthew. Go south 2 children and east 2 children. Who am I?

_____

* Start at Kate. Go west 2 children, south 2, west 2 and south 1 more. Who am I?

_____

ONE WRONG MOVE

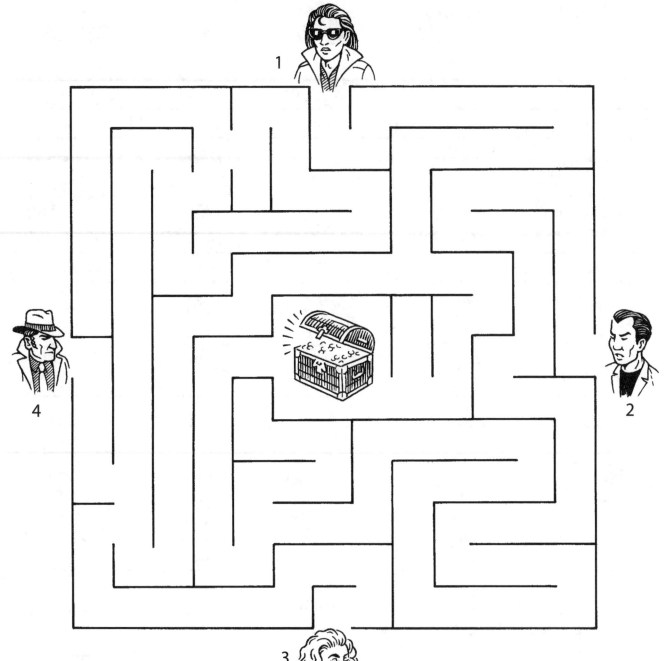

✳ Which shady character can reach the gold first and get back out of the maze? Your teacher will tell you how to get them there.

✳ Record each character's moves in the chart, using the points of the compass. (Continue on the back of the sheet if you need to.)

	Characters		
1	**2**	**3**	**4**

SANTA SPECIAL

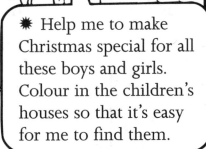

Emma	3 Holly Road	**Liam**	4 Snow Lane
James	39 Santa Street	**Luke**	51 Santa Street
Scott	24 Christmas Road	**Flora**	63 Santa Street
Dan	29 Christmas Road	**Clare**	12 Rudolf Lane
Joe	2 Ivy Street	**Karen**	17 Holly Road
Jake	45 Mount Present	**Atul**	3 Cracker Close
Becky	7 Ivy Street	**Josh**	1 Treelights Road
Ling	6 Holly Road	**Bibi**	36 Lapland Road
Kelly	29 Christmas Road	**Jamie**	40 Santa Street
Jenny	22 Ivy Street	**Cain**	6 Snow Lane

✳ Help me to make Christmas special for all these boys and girls. Colour in the children's houses so that it's easy for me to find them.

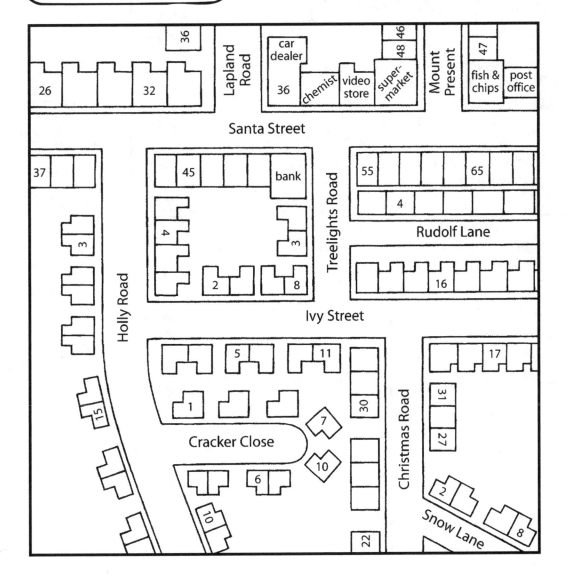

✳ Join the corners of the map together with diagonal lines. Then work out how many of the children live to the north, south, east or west.

✳ Draw a bar chart to show the results.

RAMBLING ON

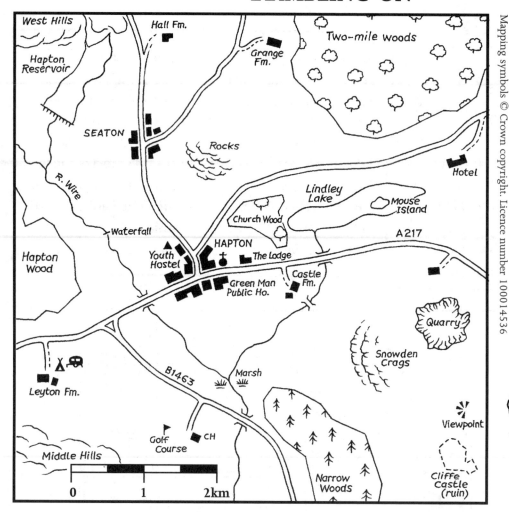

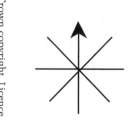

Mapping symbols © Crown copyright. Licence number 100014536

✳ Fill in the missing compass directions on the walks.

✳ Which is the longest walk?

_____

Walk 1

From The Lodge, walk _____ through Church Wood and _____ over the rocks to Seaton. Follow the road _____ to Hapton then _____ back to The Lodge.

Walk 2

Walk _____ then _____ to Castle Farm, over the bridge, then _____ to Narrow Woods. Go _____ through the woods. Walk _____ along the B1463 over the bridge. Continue to the junction with the A217 and follow this _____ to The Lodge.

Walk 3

Walk _____ along the A217 to the bridge over the river Wire, then go _____ to the waterfall. Continue _____ to the reservoir. Go _____ to Hall Farm and then _____ to Grange Farm. Walk _____, going through Church Wood to The Lodge.

VAN DRIVER

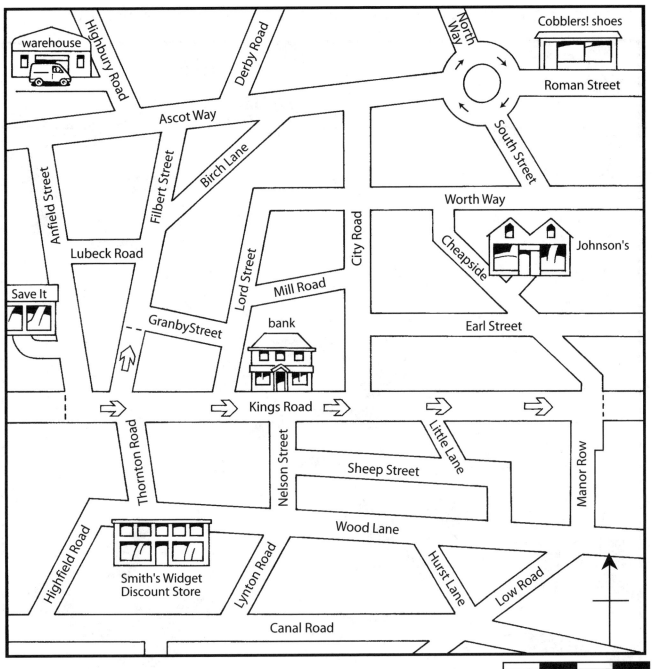

✳ Work out the shortest route to make these deliveries in order. Record your directions on a separate sheet.

1. Take 3 cartons of widgets from the warehouse to Smith's Widget Discount Store on Wood Lane.

2. From Smith's, deliver 6 boxes of chocolate teapots to Johnson's on Cheapside.

3. Go to Save It on Anfield Street and pick up 20 boxes of shoes to take to Cobblers! on Roman Street.

4. Now go to the bank on Kings Road then return to the warehouse.

COPS AND ROBBERS

✴ Try to catch the robbers. Record your move each time you throw the dice.

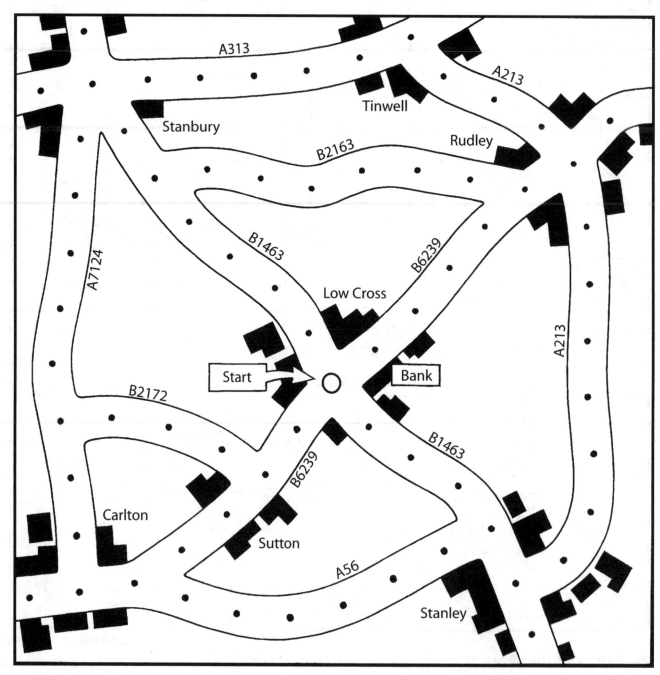

CRIMINAL PURSUIT RECORD					
Roll	Direction	Road	Roll	Direction	Road
1			8		
2			9		
3			10		
4			11		
5			12		
6			13		
7			14		

BEARINGS

✳ Use an atlas to identify and label the cities marked. The list below will help you.

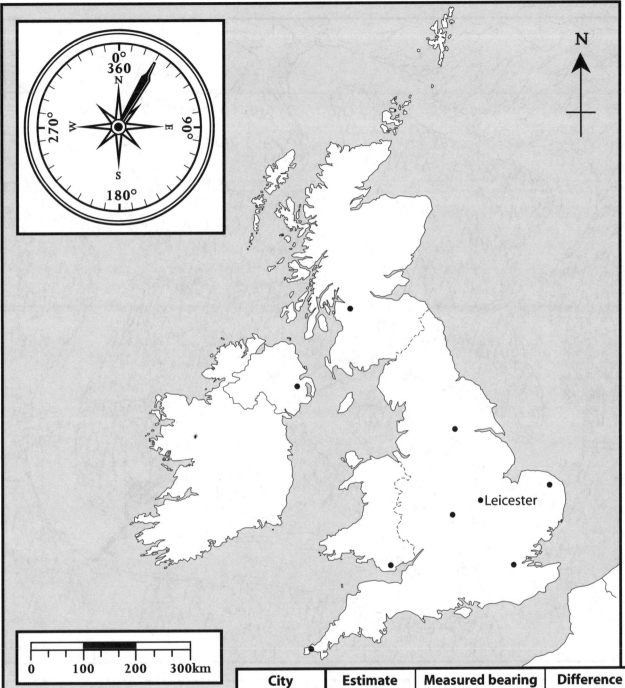

✳ Mark and label your town or village.

✳ Estimate the bearing from Leicester to each place. Then measure the actual bearing. Work out the difference between those two figures to see how accurate you were.

City	Estimate	Measured bearing	Difference
Norwich			
London			
Cardiff			
Birmingham			
Belfast			
Leeds			
Glasgow			
Penzance			

RALLY CROSS

✳ Try to win the rally. Record your estimated bearing and score each time you throw the dice.

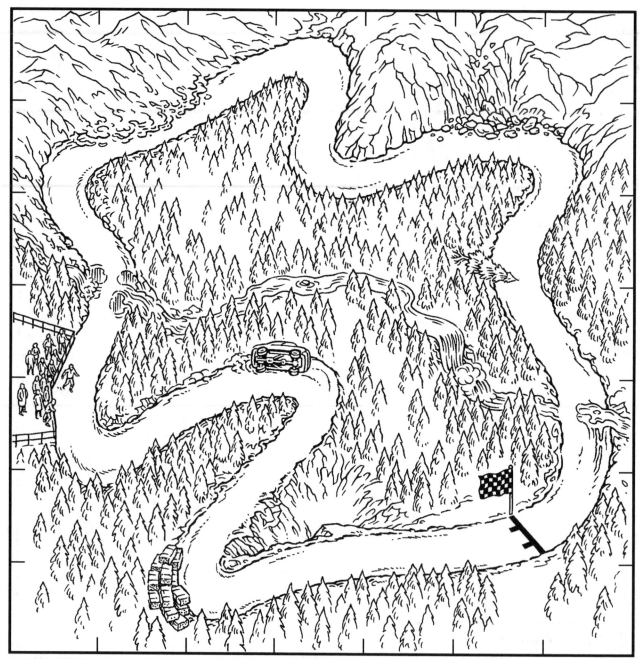

DRIVER 1				DRIVER 2			
Bearing	Score	Bearing	Score	Bearing	Score	Bearing	Score

CLAY PIGEON

✳ Aim for the centre of each clay pigeon. Estimate the bearing, then measure it.
Score **5** points if you hit the centre, **2** if you hit the edge.

1.

2.

3.

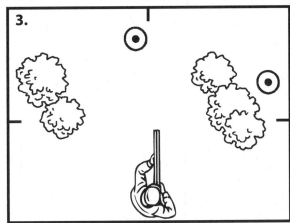

4.

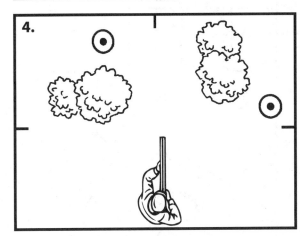

5.

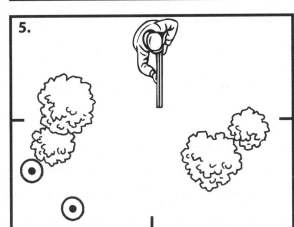

6.

	SHOOTER 1			SHOOTER 2	
	Bearing	**Points**		**Bearing**	**Points**
1			1		
2			2		
3			3		
4			4		
5			5		
6			6		
	TOTAL:			**TOTAL:**	

RÉSISTANCE

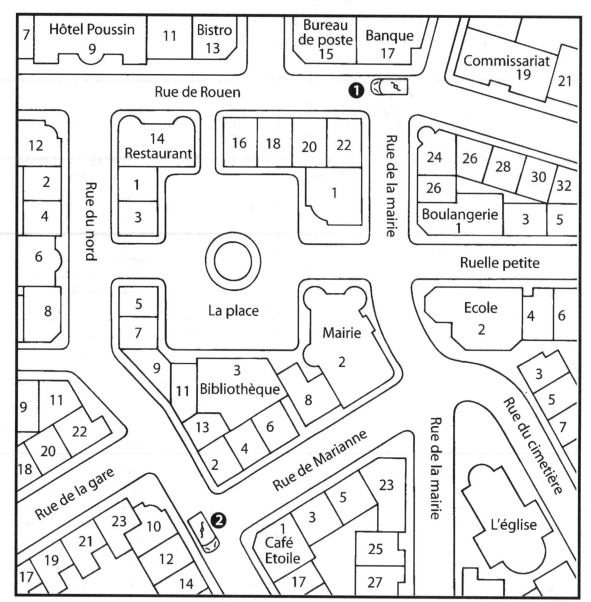

| 7 | Hôtel Poussin 9 | 11 | Bistro 13 | | Bureau de poste 15 | Banque 17 | | Commissariat 19 | 21 |

Rue de Rouen

❶

12 · 2 · 4 · 6 · 8

Rue du nord

14 Restaurant · 1 · 3

16 · 18 · 20 · 22 · 1

Rue de la mairie

24 · 26 · 28 · 30 · 32 · 26 · Boulangerie 1 · 3 · 5

Ruelle petite

La place

Mairie 2

Ecole 2 · 4 · 6

3 · 5 · 7

5 · 7 · 9 · 11 · 3 Bibliothèque · 8

9 · 11 · 22 · 20 · 18

13 · 2 · 4 · 6

Rue de Marianne

Rue de la gare

23 · 10 · 21 · 19 · 17 · 12 · 14

❷

Café Etoile 1 · 3 · 5 · 23 · 25 · 17 · 27

Rue de la mairie

Rue du cimetière

L'église

✴ Draw lines from the detector vans along the bearings.
You will find a radio transmitter where a pair of lines cross.

BEARINGS		
Van 1	Van 2	Address
210°	23°	
143°	61°	
180°	80°	
217°	13°	
231°	343°	
120°	49°	
283°	2°	
260°	335°	

LITTLE 'N' LARGE

✳ Cut out these animals and put them in order of real size. Start with the smallest.

SIZE WISE

✳ Number these objects in order of size. Start with **1** for the largest.

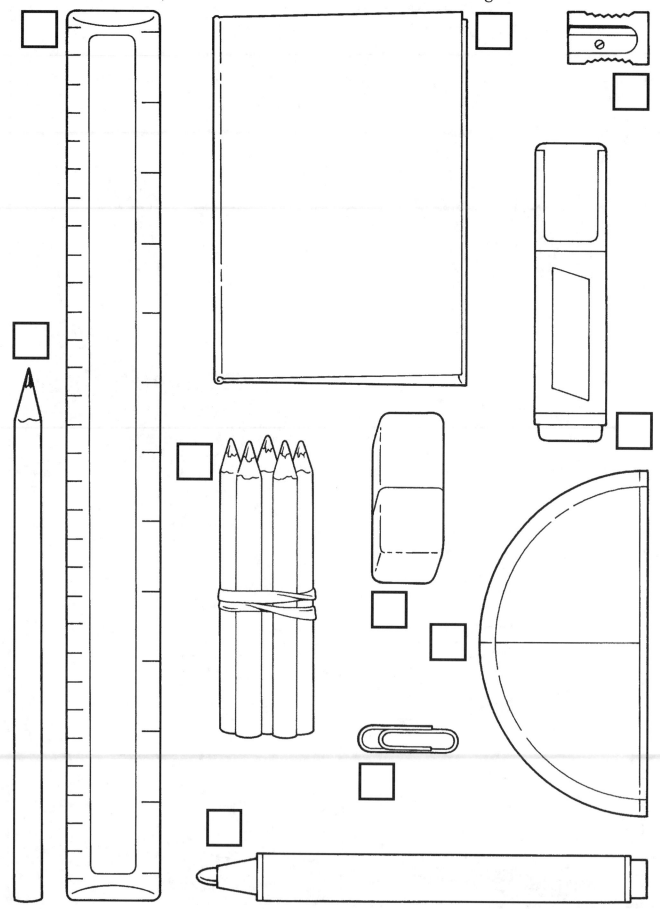

CAVENDISH STREET

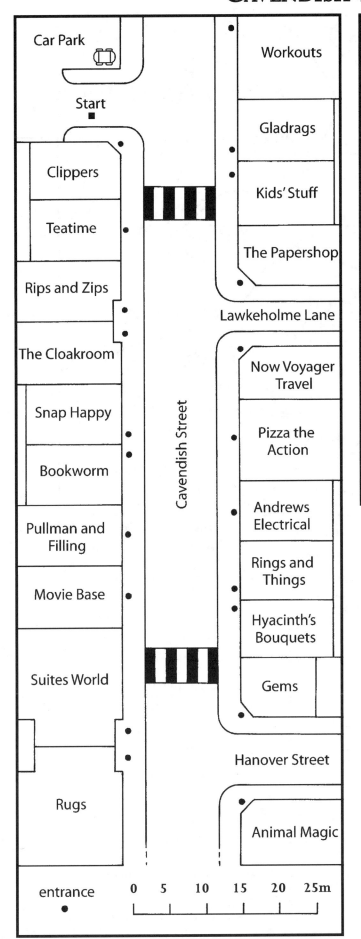

Car Park

Start

Clippers

Teatime

Rips and Zips

The Cloakroom

Snap Happy

Bookworm

Pullman and Filling

Movie Base

Suites World

Rugs

entrance

Cavendish Street

Workouts

Gladrags

Kids' Stuff

The Papershop

Lawkeholme Lane

Now Voyager Travel

Pizza the Action

Andrews Electrical

Rings and Things

Hyacinth's Bouquets

Gems

Hanover Street

Animal Magic

0   5   10   15   20   25m

SHOPPING LIST

	Distance
Take photos to be developed	_____
Haircut	_____
Plug for kettle	_____
Rent video	_____
Newspaper	_____
Flowers for Gran	_____
Dentist – check-up	_____
New dog-bowl for Portia	_____
Gym class	_____
Book holiday	_____

✳ Work out how far I walk altogether if I do everything in the order it is on my list.

_____

✳ What is the shortest distance I could have walked?

_____

DISTANCE CHART

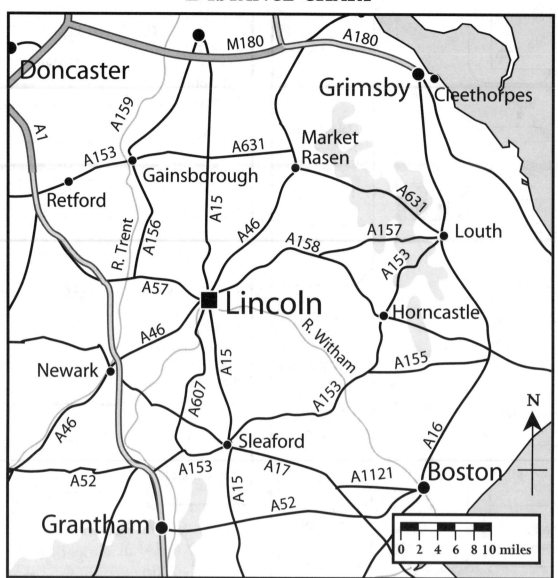

Doncaster

M180 · A180

Grimsby · Cleethorpes

A159

A1

A153

Retford

Gainsborough

A631

Market Rasen

A631

A157 · Louth

A156

A15

A46

A158

A153

R. Trent

A57

Lincoln

R. Witham

Horncastle

A153

A46

A155

Newark

A15

A607

A153

N

A46

Sleaford

A17

A16

A153

A1121

Boston

A52

A15

A52

Grantham

0 2 4 6 8 10 miles

DISTANCE CHART (miles)

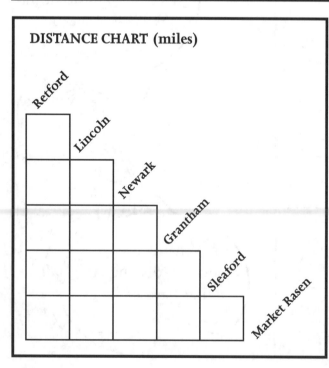

Retford

Lincoln

Newark

Grantham

Sleaford

Market Rasen

✻ Use the scale to complete the distance chart.

✻ Start at Gainsborough, where the milometer reads **14536**, travel to Lincoln, then Newark and on to Grantham. Fill in what the milometer reads in Grantham.

miles

VINCE

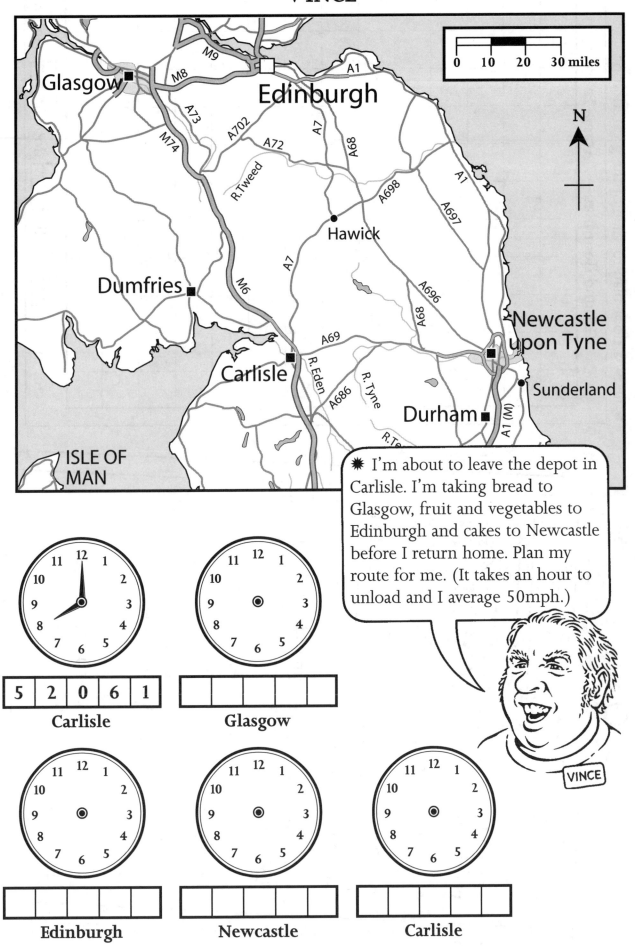

I'm about to leave the depot in Carlisle. I'm taking bread to Glasgow, fruit and vegetables to Edinburgh and cakes to Newcastle before I return home. Plan my route for me. (It takes an hour to unload and I average 50mph.)

VINCE

5	2	0	6	1

Carlisle

Glasgow

Edinburgh

Newcastle

Carlisle

❋ Draw a floor plan to scale of the ground floor.

Ground floor plan – 7 Acacia Drive

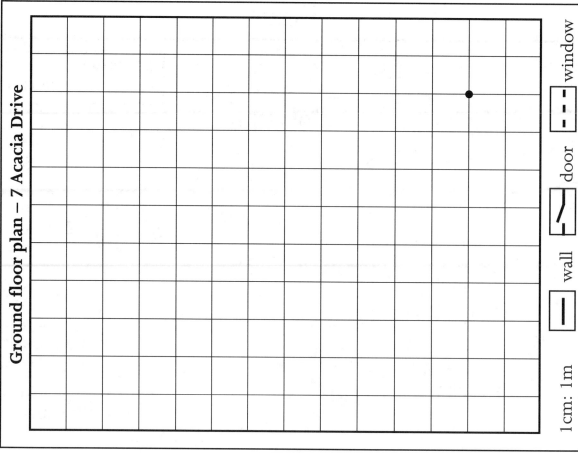

1cm: 1m

☐ wall   ⌐ door   --- window

FOR SALE

7 Acacia Drive (small detached)

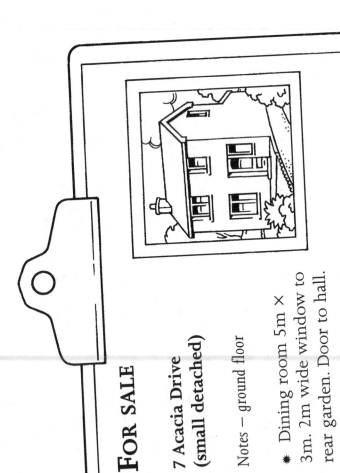

Notes – ground floor

❋ Dining room 5m × 3m. 2m wide window to rear garden. Door to hall.

❋ Living room 5m × 3m. 2m wide window to front garden. 1m wide window to side. Door to hall.

❋ Kitchen 4m × 3m. 1m wide window to rear. Door to hall.

❋ Hall 3m × 6m. Exterior doorway to front, small window to front ½m wide. Understairs cupboard. Stairs to first floor.

Good decoration. All rooms have CH. Telephone point in hall.

TERMINAL 5

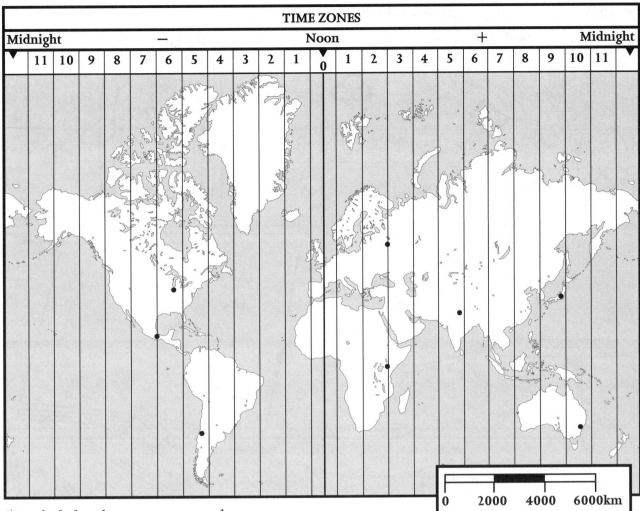

TIME ZONES

| Midnight | — | Noon | + | Midnight |

| 11 | 10 | 9 | 8 | 7 | 6 | 5 | 4 | 3 | 2 | 1 | 0 | 1 | 2 | 3 | 4 | 5 | 6 | 7 | 8 | 9 | 10 | 11 |

0   2000   4000   6000km

* Label the destinations on the map.
* Complete the departures board.

HEATHROW	DEPARTURES		TERMINAL 5
Destination	Direction	Distance (km)	Flight time
Chicago			
Mexico City			
Tokyo			
Sydney			
New Delhi			
Santiago			
Nairobi			
Moscow			

WHAT'S MISSING?

✸ Draw the missing symbols in the correct boxes.

PUFFIN ISLAND

❋ All these people run a post office on Puffin Island. Invent a symbol to represent a post office. Put it in the key and add it to the map to show where the post offices are.

POST OFFICE
Mr Thompson, Helmton
Mrs McPhee, Portby
Mr Armstrong, Barrowby

❋ Now do the same for these people. Complete the key before adding your symbols to the map.

GARAGE
Mr Fixit, Helmton

SCHOOL
Mrs Masters, Helmton

PUB
Mr Bass, North Cape
Mrs Porter, Marun
Mr Peacock, Barrowby
Mr Marston, Skipsea

CHEMIST
Mr Potion, Helmton
Mrs Fine, Portby

TAXI SERVICE
Mr Driver, Portby
Miss McHine, Skipsea

NEWSAGENT
Mrs Lane, Helmton
Miss Moore, Barrowby

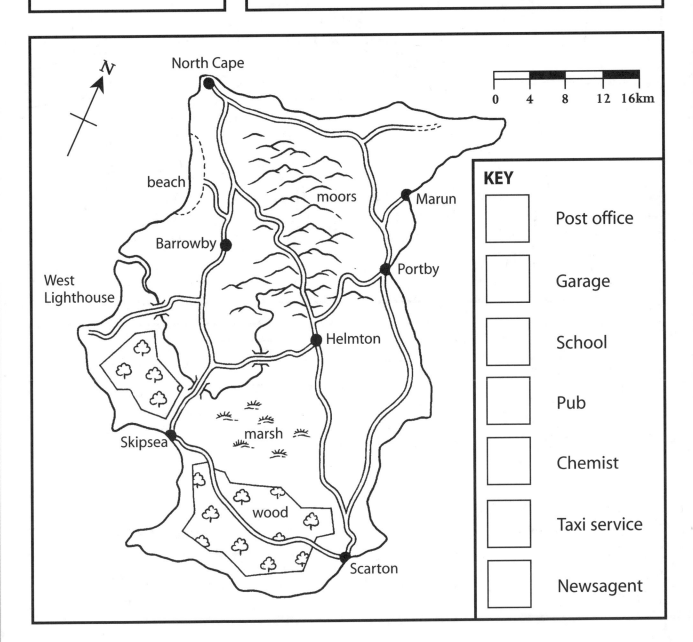

KEY

☐ Post office

☐ Garage

☐ School

☐ Pub

☐ Chemist

☐ Taxi service

☐ Newsagent

EASTWICK FARM

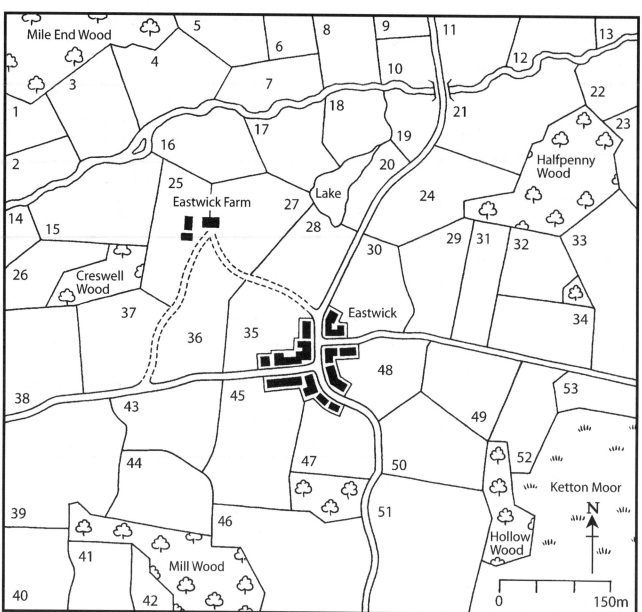

* Complete the key, then use it to make a colour-coded, land-use map of Eastwick Farm.

KEY

Moorland	
Woodland	
Lake	
	Sheep 47, 49–53

	Cattle 1–4, 7–22
	Wheat 25, 27–29, 33, 37–39, 43, 45, 48
	Barley 5, 6, 23–24, 31–32, 34, 40
	Oats 26, 41–42
	Oil-seed rape 30, 35–36, 44, 46

LAXTON

* Complete the key and colour the map.
* Draw lines to link the features on the left to their positions on the map.
* In the boxes on the right, draw the features listed.

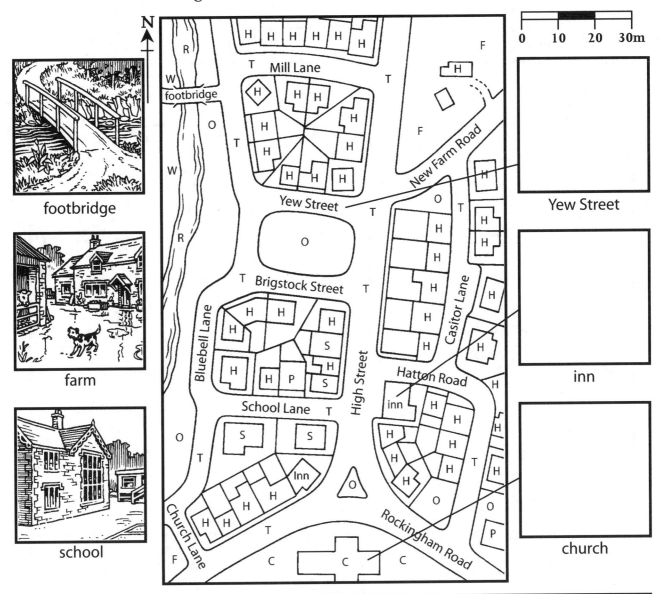

footbridge

farm

school

Yew Street

inn

church

KEY									
Colour	Letter	Land use	Rank		Colour	Letter	Land use	Rank
	W	Woodland				F		
	R						House	
		Open space				S		
		Church					Public buildings	
		Transport						

TOURIST INFORMATION

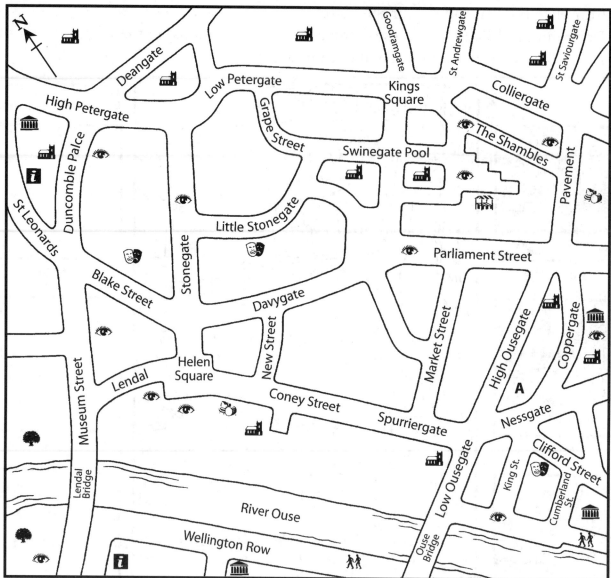

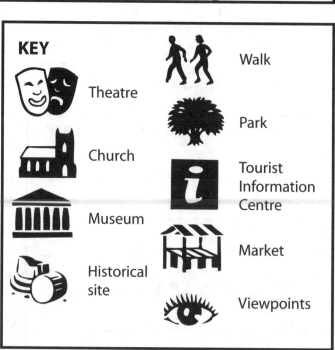

KEY

🎭 Theatre

⛪ Church

🏛 Museum

🪙 Historical site

🚶 Walk

🌳 Park

ℹ️ Tourist Information Centre

🎪 Market

👁 Viewpoints

✳ Use the map and key to answer these questions. (Write your answers on a separate sheet.)

1. What places of interest are there near St Leonards and Duncomble Place?

2. How many churches are there in this part of York?

3. What places of interest are there within 150m of point **A**?

4. Describe a short tourist bus journey around York.

SYMBOLS HUNT

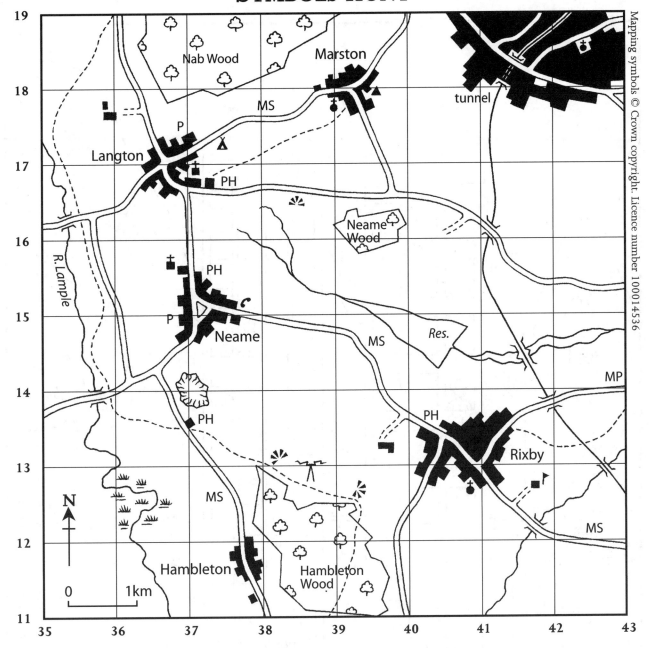

※ On a separate sheet of paper, compile a key to explain all the symbols on the map, for example ♣ for a church with a tower and ⌣ for a bridge. Then answer these questions.

1. What lies at 404136?

2. What will you find at 412161?

3. Put a church with a tower at 377116.

4. Put a bridge at 427120.

5. Add a milestone at 415156.

6. Could you pitch a tent at 375172?

7. What game might you play at 418128?

8. What footwear would you need at 362124?

HIGH HOPES

❋ Members of the Hope family live, work and have fun in different kinds of places. Work out how high each one is. Use the scales to help.

Name	m	Name	m	Name	m
Simon		Sarah		Mum	
Dad		Rover		Jack	
Flo		Danny		Tom	
Amy		Tim		Kate	

◗◖ S C H O L A S T I C PHOTOCOPIABLE

Highway Code

✳ Label the signs with a suitable hill gradient from below.

HILL 'N' DALE

✳ Draw a line to join each photograph to its lettered point on the map.

✳ Try to imagine where the photographer was standing each time and draw an arrow on the map towards the view being photographed.

✳ Link all the places you have identified to make a walk around the countryside and write a description of the walk.

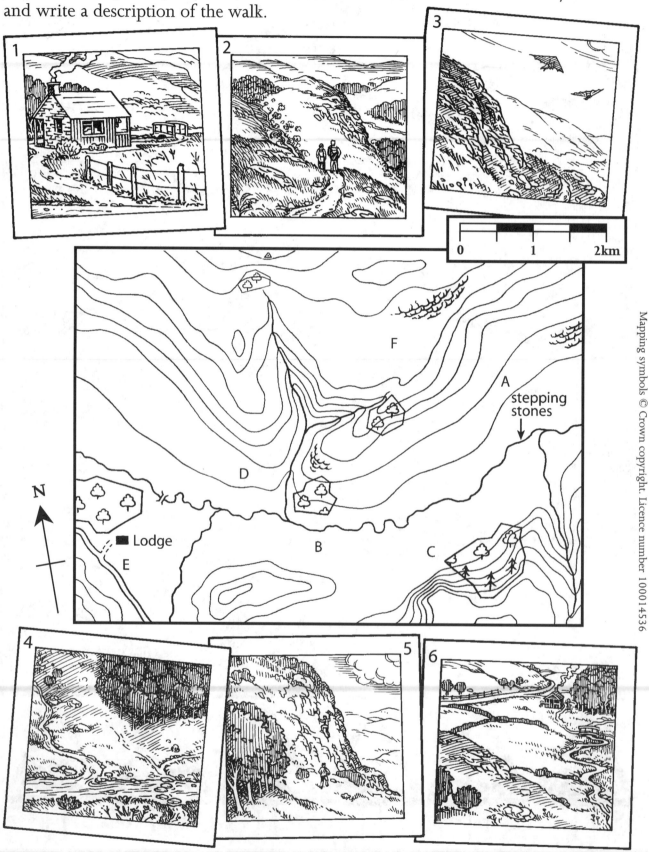

UP AND AWAY

KEY

Colour	Height
	0–50m
	50–100m

✱ Complete the key and use it to colour the map according to the height of the land.

✱ Find the highest and lowest areas of land.

✱ Draw a line across the map and describe a journey the hot air balloon would make along it.

0   1   2km

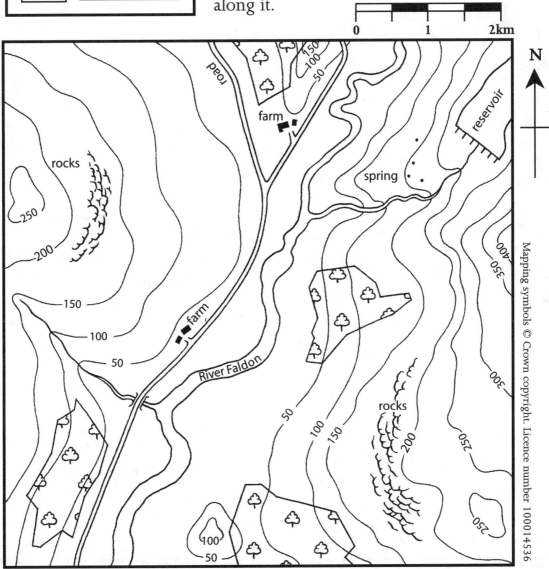

N

Mapping symbols © Crown copyright. Licence number 100014536

HUFF AND PUFF

✳ Draw and label a map of the story.

Once upon a time there were three little pigs who wanted to live in their own houses, away from a nasty wolf who was trying to eat them. So they left their parents' house to live by themselves. They went over the bridge that crossed a river onto farmland, and built a house of straw for Cedric, the smallest pig. Then they followed a path from the field and went

further away into a dark mysterious wood where they built a house for Gordon, the middle-sized pig, out of twigs. The largest pig, Tony, didn't want his house to be in a field or in the wood, so the brothers went back across the river, over another bridge, through fields scattered with trees to a hill where they built his house of bricks. The rest, as they say, is history.

LOST

✳ Read the directions given below. Use the information to draw picture maps to help the drivers get to the cinema and the school.

Excuse me. I'm lost. How can I get to the cinema please?

Go down this road, past the supermarket, turn left, go under a bridge and it's in front of you.

Excuse me. I'm lost. Can you tell me where the school is please?

Drive along by the park, turn right at the traffic lights, past a church and the bus station. It's the next building on your left.

WILD WEST ADVENTURE

✱ Help to design a Wild West Adventure theme park. Cut out the attractions and place them in the park. When you are happy with their positions, stick them down. Add a train track, and paths and a car park if you like.

✱ Draw a map of your Wild West Adventure park on a separate sheet.

Doc Holliday's first aid post

Saloon

County jail

Bucking bronco ride

Wishbones fast food

Stampede

Iron horse

Rolling logs

ROCKWELL

❋ Make a map of this part of Rockwell.
❋ Give the roads names and label the buildings.

MAP-MAKING

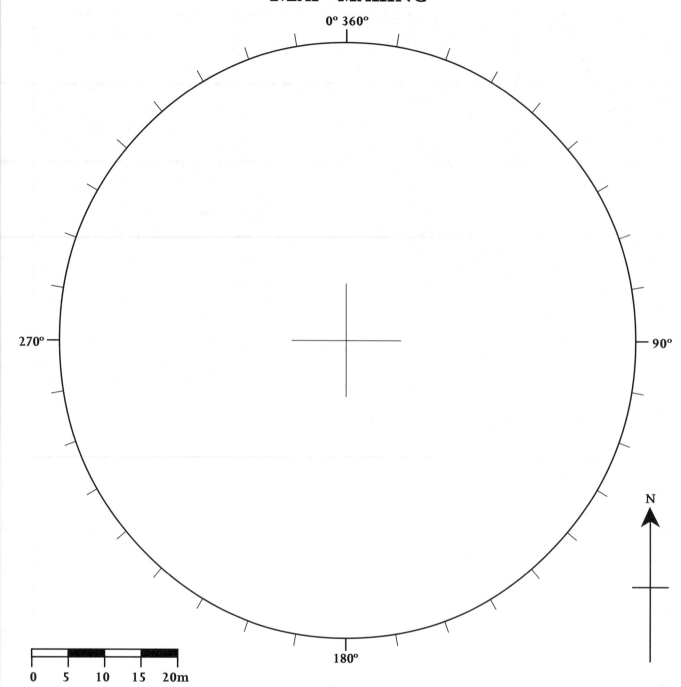

0° 360°

270°   90°

N

180°

0   5   10   15   20m

✳ To begin making a map of an area around these children's school, use the circle to put:

1. a bush at 120°, 5m away

2. a tree at 290°, 10m away

3. a litter bin at 25°, 15m away

4. a manhole cover at 195°, 13m away

5. a milk crate at 252°, 7m away.

✳ Now draw the map on a separate sheet of paper.

✳ Add some more features to the map. Keep a record on the circle of where and how far away they are.

◨ S C H O L A S T I C **PHOTOCOPIABLE**

SCHOOL CHASE (1)

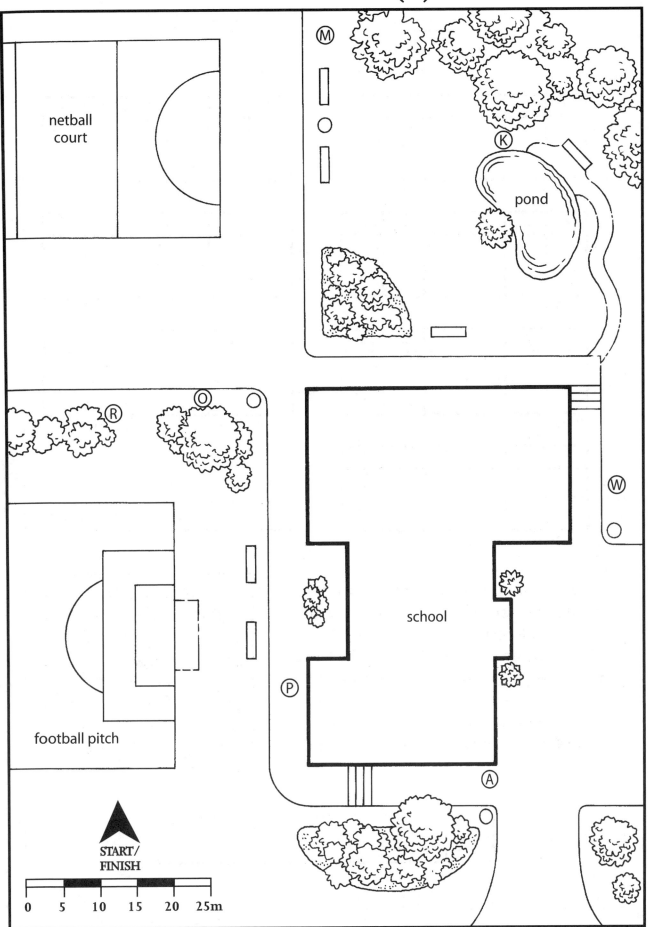

netball court

pond

school

football pitch

START/ FINISH

| 0 | 5 | 10 | 15 | 20 | 25m |

SCHOOL CHASE (2)

	B	D	B	D	B	D
P1						
P2						
	Try 1		Try 2		Try 3	

	B	D	B	D	B	D
P1						
P2						
	Try 1		Try 2		Try 3	

	B	D	B	D	B	D
P1						
P2						
	Try 1		Try 2		Try 3	

	B	D	B	D	B	D
P1						
P2						
	Try 1		Try 2		Try 3	

	B	D	B	D	B	D
P1						
P2						
	Try 1		Try 2		Try 3	

	B	D	B	D	B	D
P1						
P2						
	Try 1		Try 2		Try 3	

	B	D	B	D	B	D
P1						
P2						
	Try 1		Try 2		Try 3	

KEY B: bearing D: direction P: player

✳ Take it in turns to estimate the bearing and distance to each letter. Have up to three turns each to get inside the letter circle. When you are both there, collect the letter.
✳ As you play, draw your route around the school.
✳ You can collect the letters in any order, but you should make a word out of them at the end.

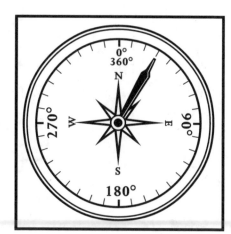

Word: [ ][ ][ ][ ][ ][ ][ ]